THE
DIGESTIVE SYSTEM
A Tour Through Your Guts

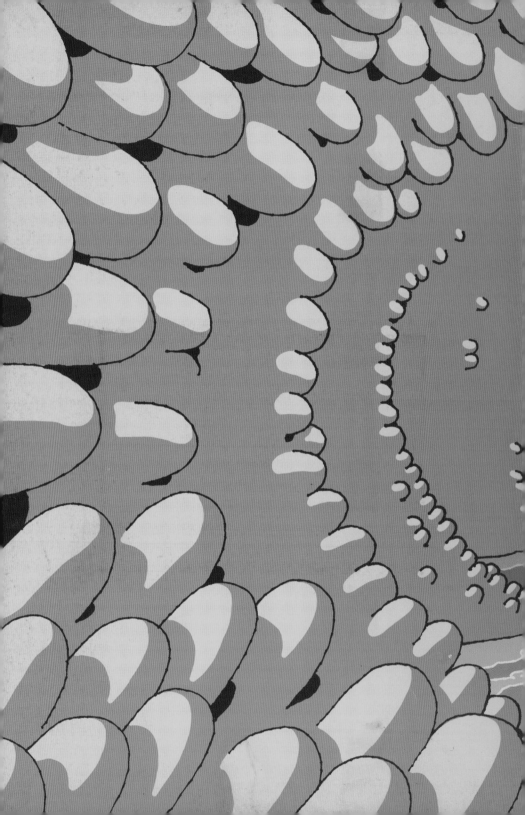

THE DIGESTIVE SYSTEM
A Tour Through Your Guts

written by
JASON VIOLA

art by
ANDY RISTAINO

:01
First Second
New York

Special thanks to John F. Stanievich, MD.
—Jason

To all the good bacteria inside us:
We couldn't have digested without you.
—Andy

First Second

Text copyright © 2021 by Jason Viola
Illustrations copyright © 2021 by Andy Ristaino

Published by First Second
First Second is an imprint of Roaring Brook Press,
a division of Holtzbrinck Publishing Holdings Limited Partnership
120 Broadway, New York, NY 10271

Don't miss your next favorite book from First Second! For the latest updates
go to firstsecondnewsletter.com and sign up for our enewsletter.

Library of Congress Control Number: 2019948173

Paperback ISBN: 978-1-250-20404-2
Hardcover ISBN: 978-1-250-20405-9

Our books may be purchased in bulk for promotional, educational, or business use. Please
contact your local bookseller or the Macmillan Corporate and Premium Sales Department
at (800) 221-7945 ext. 5442 or by email at MacmillanSpecialMarkets@macmillan.com.

First edition, 2021
Edited by Dave Roman
Cover and interior book design by Molly Johanson
Biology consultant: Brandon E. Jackson, PhD.

Printed in China by Toppan Leefung Printing Ltd., Dongguan City, Guangdong Province

Sketched and arranged in Photoshop, printed with blueline on 20-lb. legal paper,
inked with Pigma Micron 01 pens, scanned, and digitally colored in Photoshop.

Paperback: 10 9 8 7 6 5 4 3 2 1
Hardcover: 10 9 8 7 6 5 4 3 2 1

I've always wanted to be an explorer. Growing up I used to stare at maps, wondering where rivers began or what lay on the other side of the ocean. I dreamed of finding a mysterious island that no one else had ever discovered. In my imagination, I invented what kind of people might live there. I thought of the strange trees and animals that I might find. My heroes were other explorers: Magellan, Charles Darwin, Amelia Earhart. What would be more exciting, I thought, than sailing off into the unknown to discover something totally new?

As I grew up, I found it wasn't easy to be an explorer—not the kind that goes off in ships anyhow. After all, we know the names of all the islands and the seas that surround them. Still, I didn't give up. There are plenty of new places that remain unknown and even unnamed. And I can tell you I am an explorer today! I set sail for new worlds every morning. Where are these places I'm going?

Almost every day, I voyage into the darkness of the gut. I am a special kind of doctor called a gastroenterologist—or, if you want to call me this, a gastronaut. Much like Earth, the human body is a tremendous and complicated world. The gastrointestinal system is just one part of it—but it's the one that I find the most fascinating. Just as we once needed sailors to examine the unknown seas, we need explorers to study the body. We have maps, but they are incomplete. We know of places with unusual names—the villi of the jejunum, the islets of Langerhans, the sphincter of Oddi—which are as mysterious and exotic as any island in the deepest sea. And these days, we are lucky to have tools to travel there.

The ship on which I sail is called an endoscope—a highly technical device with a camera at its end—that lets me journey into the body. On a screen, I guide the camera past the shining white teeth, over the pink clouds of the tongue, and down into the dark tunnels that lead to the bright cavern of the stomach. There, I look for diseases. I catch bacteria and parasites and put them into bottles for identification. I encounter unfamiliar structures that I've never before seen. Many of the endoscopies, or procedures, I perform are on people who are sick. The purpose of such journeys is to help patients get better. During these explorations, I try to come back with answers. I can tell them I've discovered a bug that is making them ill. I can show them photographs of ulcers that we found in their small intestine so that they can be treated with medications. We can find and remove tumors that might have otherwise turned into a bad cancer.

And sometimes, these journeys are to discover something entirely new. Something no one has ever found. There's so much about the gut we don't know, and part of my job is to research what is unknown or uncharted. The deepest part of the intestines are full of millions of creatures—unknown micro-organisms, viruses, bacteria—that still need to be named and understood. We believe these organisms play a key role in our health, but how they do this is hardly clear. And there's more. The walls of the intestines are full of dense forests of nerve endings that hold secrets about how we feel sensations like pain and even emotions such as happiness. The cells of the stomach have something to say about how we use sugar and protein and fats, and even how our body might help regulate its internal clock—the sense of time. The liver is almost as mysterious to us now as it was a century ago, with new discoveries about how it works being made every day with each journey one of us takes.

Sometimes, after I'm done with work, I go for a run. If I go far enough, I reach the harbor and can look out to the horizon. Even now, I have the same feelings as I did when I read about sailors long ago. There's nothing in this world, I think to myself, that is more exciting than a journey to someplace new. And that journey can be inward, as much as it was once a voyage into the sea. In this book, we wanted to share with you some of that mystery and excitement as we travel deep inside the body.

Will you join us?

Sushrut Jangi, MD
gastroenterologist

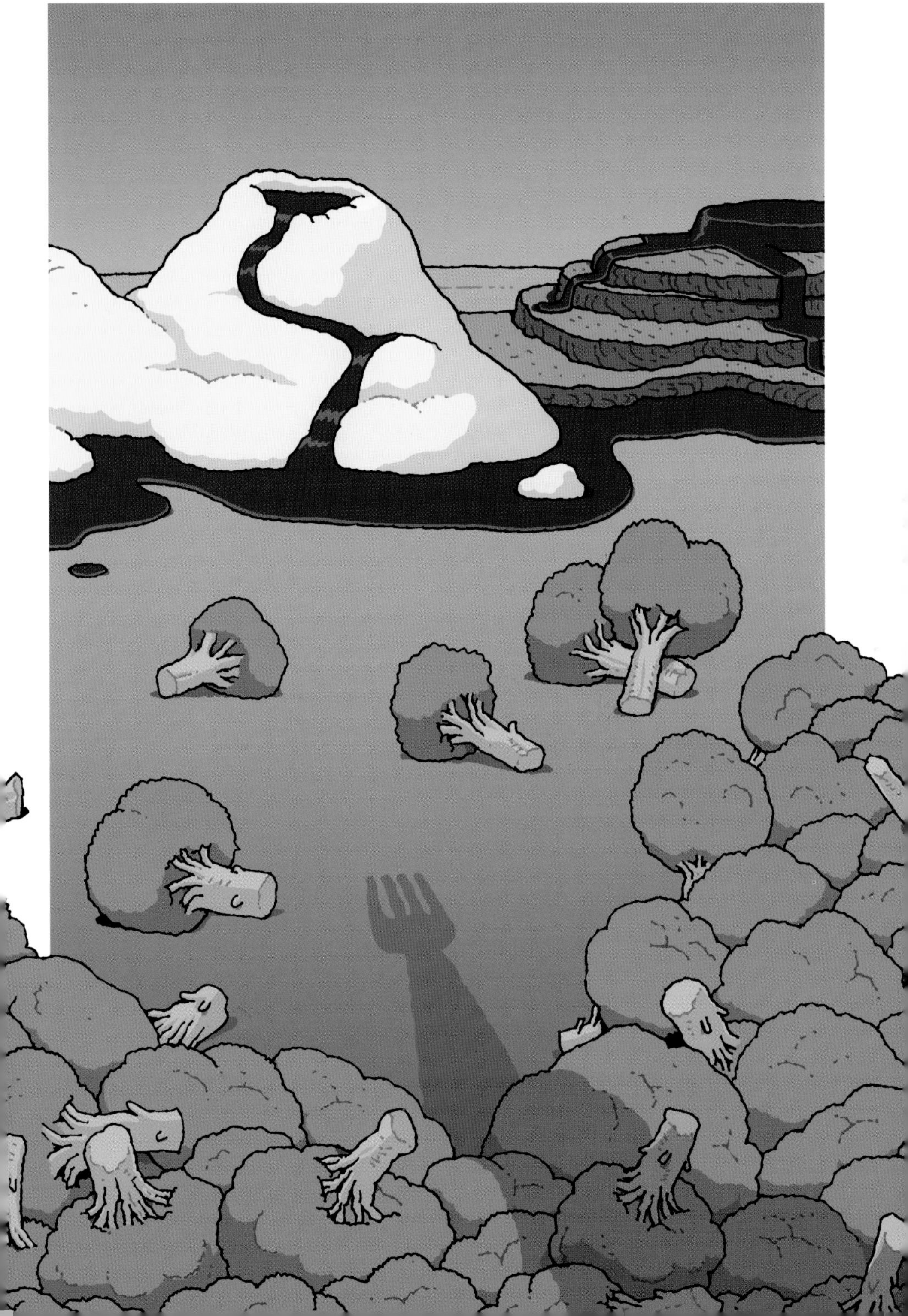

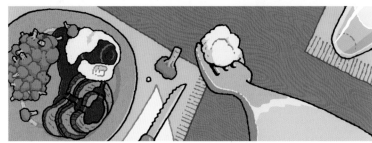

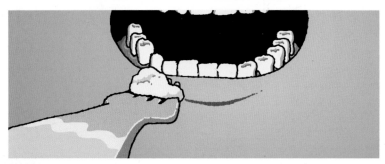

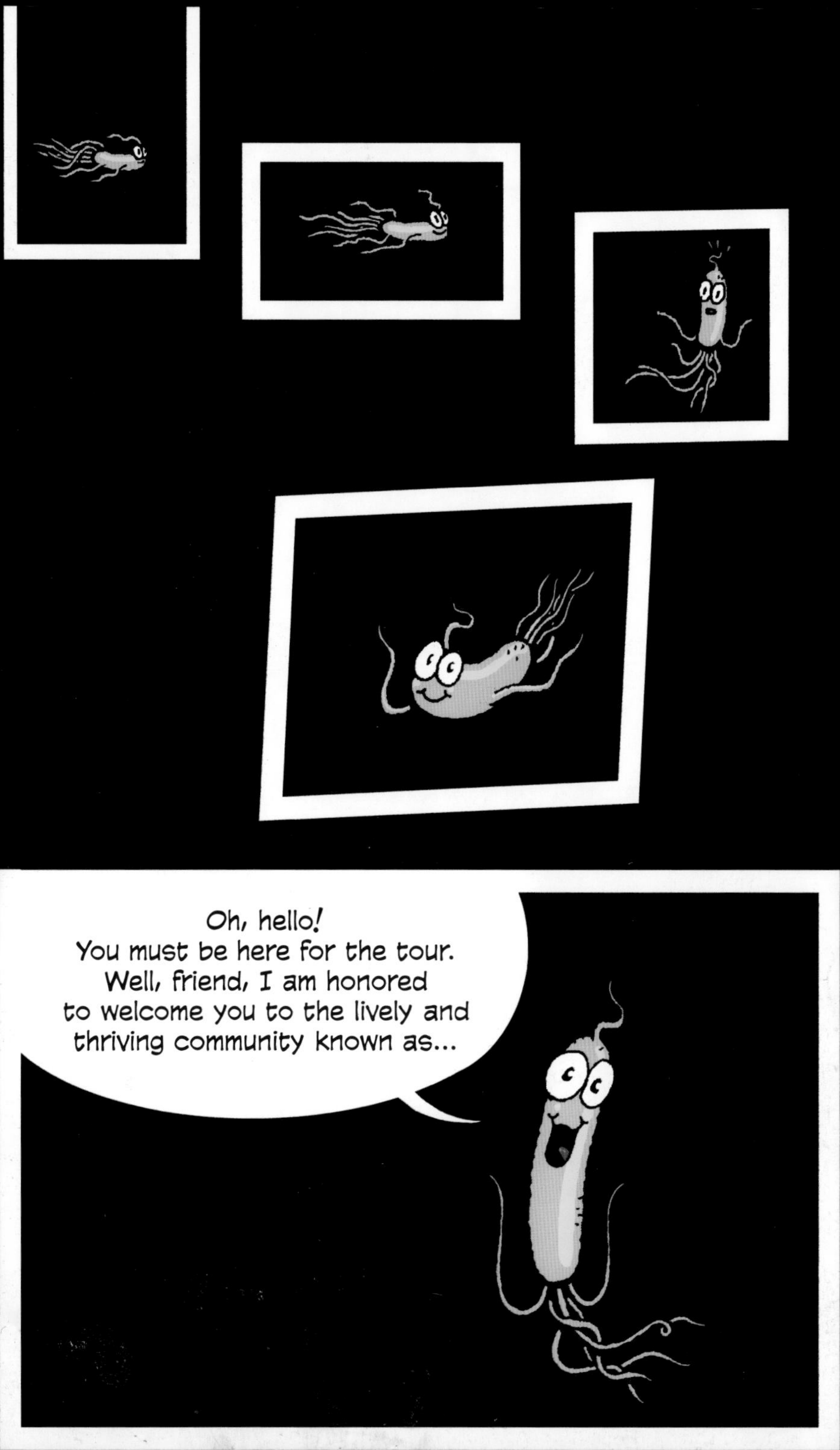
Oh, hello!
You must be here for the tour.
Well, friend, I am honored
to welcome you to the lively and
thriving community known as...

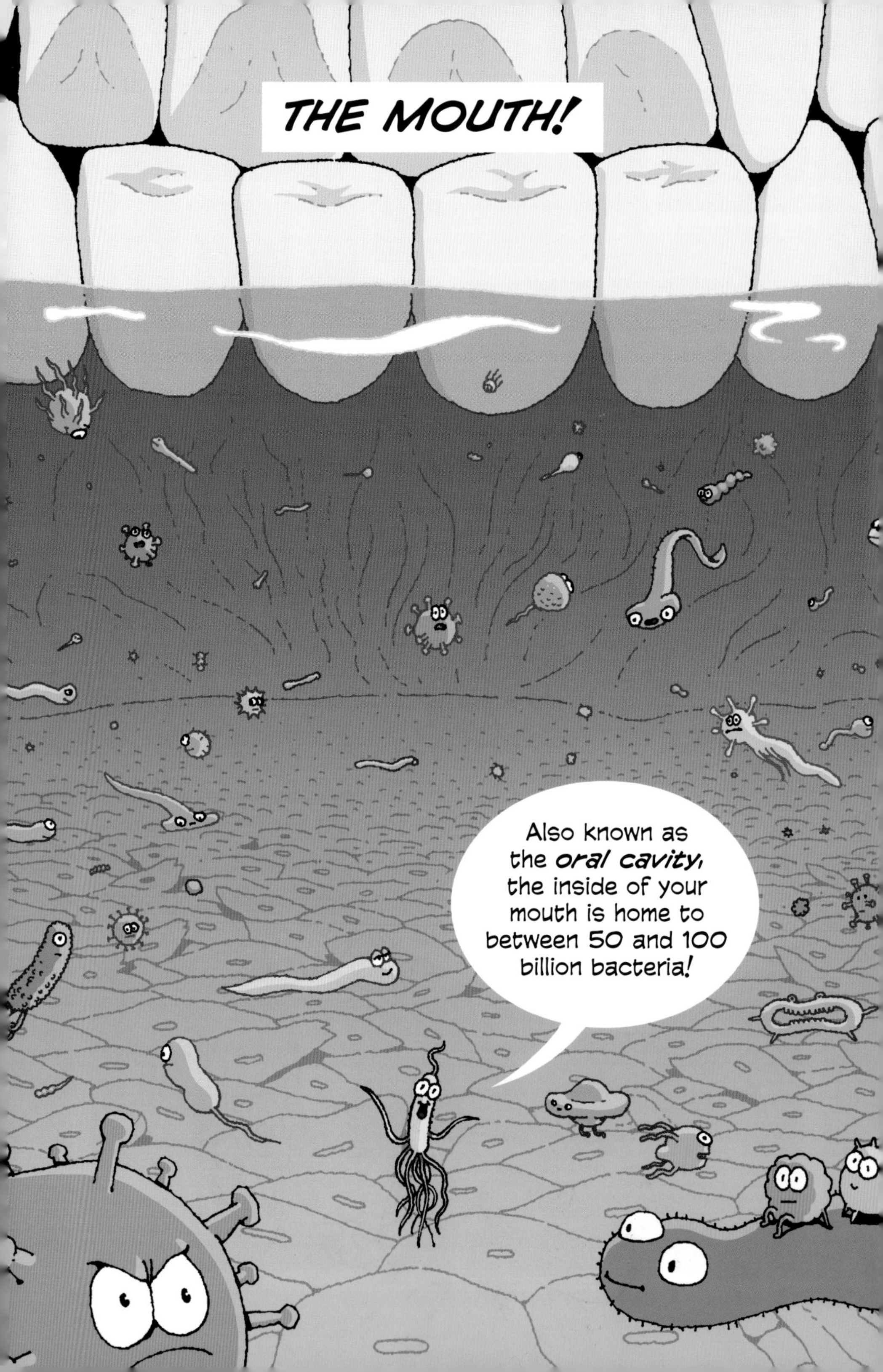
THE MOUTH!
Also known as the *oral cavity*, the inside of your mouth is home to between 50 and 100 billion bacteria!

You're lucky you found me first because there are about **200 species** of bacteria living here at any given time, and some of them are impolite.

Ugh, who invited the noob?

Just ignore them for now.

My name's E, and I'm happy to guide you around!

You may be surprised to learn that the mouth is not just a booming microbial metropolis— it also houses sensitive and powerful organs that are used for *digestion*.

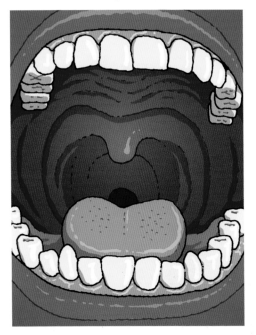

Digestion is the process of taking complex food...

...and breaking it down...

...into simple, basic forms.

I'm telling Mom!

You see, all living things need *energy* to grow and move.

That includes the plants and animals that humans use for food.

Food is a mixture of chemical substances called *nutrients*.

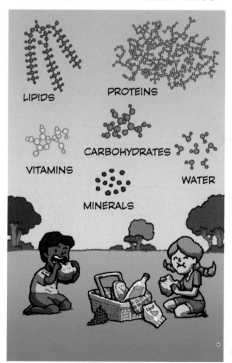

LIPIDS

PROTEINS

VITAMINS

CARBOHYDRATES

WATER

MINERALS

Through physical and chemical means, digestion loosens and separates the nutrient molecules so they can be absorbed by your body's cells.

Farewell, fructose! Never forget what we had.

Metabolism is the process of transforming food into energy and using that energy to grow and sustain life.

Oh, uh...
hang on just a
second...

Ungh! Sorry, this happens
from time to time!

Hi! Welcome to...
THE MOUTH!

POP

I already told
them that.

Oh.
Where did
you leave
off?

I just started telling
them about nutrients.

Right,
nutrients!

You could tell them
about the different
types of—

There are two types of
nutrients: macronutrients
and micronutrients!
Macronutrients are
needed every day in
large amounts.

MACRONUTRIENTS

1. Carbohydrates
2. Proteins
3. Lipids

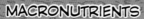

Wow, you got
that text to appear
super fast...

Carbohydrates, also known as sugars, provide the body with fuel!

Fill 'er up with glucose today.

Single or double molecules of carbohydrates are called simple sugars. *Glucose*, a simple sugar found in all fruits and vegetables, is the most essential source of energy in the body.

It's all I ever order!

More than three sugar molecules in a chain are called "complex" carbohydrates, and no longer taste sweet. Breaking them back down releases all the simple sugar molecules that had been bound together.

Plants make starches to store energy that they don't need right away. They can also use glucose to make *fiber*. Fiber is a complex carbohydrate that supports their leaves and stems.

You're not that complex, Bradley.

Got another batch of cellulose for those cell walls!

Like complex carbohydrates, *proteins* are chains of smaller units. Twenty types of *amino acids* can be recombined to make different proteins. Nine of them can be converted into the other eleven. But those nine "essential" amino acids must be supplied by food directly. There are a few "complete" proteins, which have all the essential amino acids.

Cuisines have evolved to combine "incomplete" proteins, which together provide the amino acids needed by the body. Some examples are rice and beans, pasta and cheese, pita bread and hummus, black beans and tortillas, soba noodles and peanut sauce!

While carbohydrates provide energy for your cells, proteins are their building materials! Protein is needed to develop and maintain your muscles, organs, blood, and bones. Proteins called *enzymes* carry out all chemical reactions. And they build all other molecules, including those complex carbohydrates and even other proteins!

You want a heart? We can build you a heart.

Lipids, such as fats, are found in food like meat, oil, nuts, and eggs. They can be used to protect your organs, make hormones, and coat the outside of your nerves.

Most of the lipids found in food are *triglycerides*, which provide energy to muscle cells at rest! When your body has extra energy, it makes its own triglycerides to store it and to protect organs.

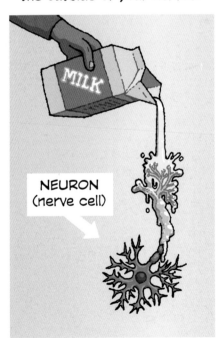

NEURON
(nerve cell)

Your coat looks so warm and protective, Cynthia.

It's 100% triglyceride!

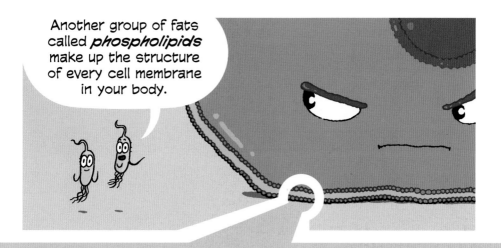

Another group of fats called *phospholipids* make up the structure of every cell membrane in your body.

Their head is drawn toward water while their two tails are attracted to fat. This helps shuttle other lipids out of your watery cells.

TAIL

PHOSPHOLIPID BILAYER

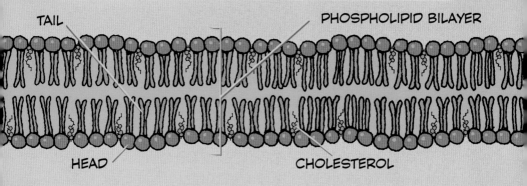

HEAD

CHOLESTEROL

Lipids in the form of *cholesterol* act as a buffer, both spacing out the phospholipids to let substances through while also preventing too large of a gap from forming between them.

Thanks for nothing, big guy!

Get lost, bacteria.

The human body also needs *minerals*. Unlike other nutrients, minerals aren't manufactured by any living things; they are elements that come from the earth.

The five minerals that are needed in large amounts do a lot of different things! Here are some functions they help with:

Calcium - strengthens bones and teeth

Magnesium - used in bone formation, muscle and nerve function, immune system

Phosphorus - assists in protein production and energy storage

Potassium - supports nerve function and muscle contraction

Sodium - regulates nerves and muscles

Micronutrients are needed in small amounts. They include trace minerals like iron and zinc, as well as vitamins. Vitamins are needed for the chemical reactions that convert food to energy.

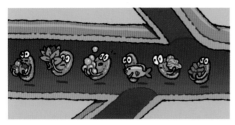

Some vitamins are *water soluble*, which means they can be dissolved in water. They are absorbed into the bloodstream, where your cells can hit them up directly! But they can be lost in urine, so they must be replenished daily.

Other vitamins are *fat soluble*. Instead of water, they dissolve in fat and can be stored there for long periods of time.

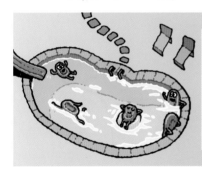

Nearly all food contains *water*, a chief component of all your body's cells, and is essential for life. 55–60% of an adult's body weight is water!

It transports nutrients...

...regulates body temperature...

...and carries away waste.

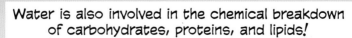

Water is also involved in the chemical breakdown of carbohydrates, proteins, and lipids!

That was fantastic— thank you so much!

My pleasure!

You should stick around for the rest. Think about it— two bacteria covering the entire mouth!

Sorry, looks like my ride's coming. Gotta bounce!

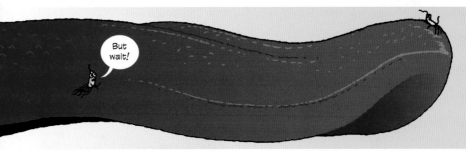

But wait!

They never stay.

But that's okay! You've got a trustworthy guide in me. Look, I even drew this map of the mouth for visitors like you to get acquainted with the area.

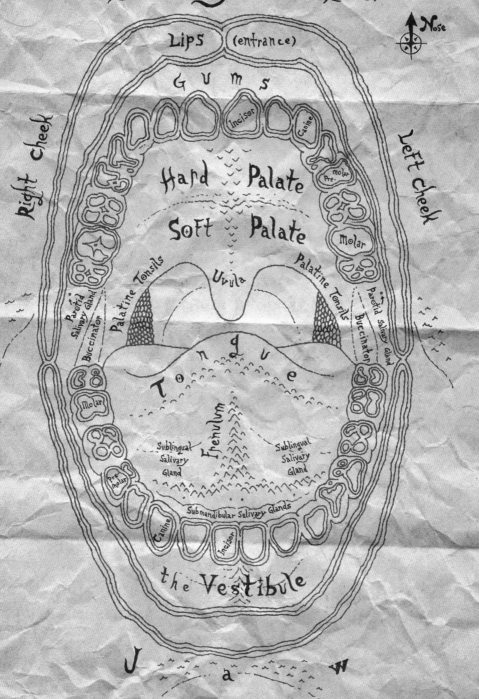

Oral Cavity & Environs.

Nose

Lips (entrance)

GUMS

Right cheek

Left cheek

Incisor

Canine

Hard Palate

Pre-molar

Soft Palate

Molar

Palatine Tonsils

Uvula

Palatine Tonsils

Parotid Salivary Gland

Buccinator

Palatine Tonsils

Parotid Salivary Gland

Buccinator

Tongue

Molar

Frenulum

Sublingual Salivary Gland

Sublingual Salivary Gland

Pre-molar

Canine

Submandibular Salivary Glands

Incisor

the Vestibule

J a W

Now, when food enters the mouth, it can be physically broken down by chewing it. This process is called *mastication*. Up to the task are your jaws, whose muscles are the most powerful in the body.

When you chew, your jaws put forth a pressure of up to 80 kilograms (180 pounds), about the weight of a grown man! Your teeth working together can become quite a force.

But it's not force alone that make the jaws so powerful...

...it's their sensitivity and ability to protect. If not for your jaw muscles' keen detection, your teeth would bash each other to their *pulps!* Instead, they step on the brakes the moment they sense that your food has given way. The faster you chew, the less force your muscles use; they can't risk a slipup!

The tissue that makes up the visible part of a tooth is *enamel*, the hardest substance produced by the body! And inside the tooth is the *pulp*, which contains nerves and blood vessels that exit through a hole in the root.

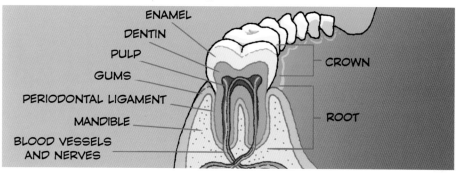

ENAMEL

DENTIN

PULP

GUMS

PERIODONTAL LIGAMENT

MANDIBLE

BLOOD VESSELS AND NERVES

CROWN

ROOT

Cross me once and I will *cut* right into you. This I promise.

I can cut and rip if I must, but *crushing* is my game. Then I pass to Molar for grinding!

ME IS CANINE! ME GRAB, RIP, AND TEAR!

INCISOR!

PreMolar!

CANINE!

MOLAR!

I'm the finisher.

I'd like to introduce you to my pals, the mutans streptococci! We are hanging out on the surface of the teeth, where there are over 600 different microorganisms that form a mass called *dental plaque!*

These bacteria love sugar! When they digest sugars like glucose, they produce acid, which contributes to tooth decay and gum disease!

We aren't exactly pals.

This is literally the first time we've met.

Hey, enough with the propaganda! We can't control ourselves.

If people don't like it, they should brush and floss their teeth.

I thought we were pals!

Next stop: *THE TONGUE!*

Who was that jerk?

Anchored to the floor of the oral cavity is the tongue, made of muscle tissue that allows for expansion, compression, and movement in nearly every direction. This incredible flexibility helps you guide food around and keep it in play.

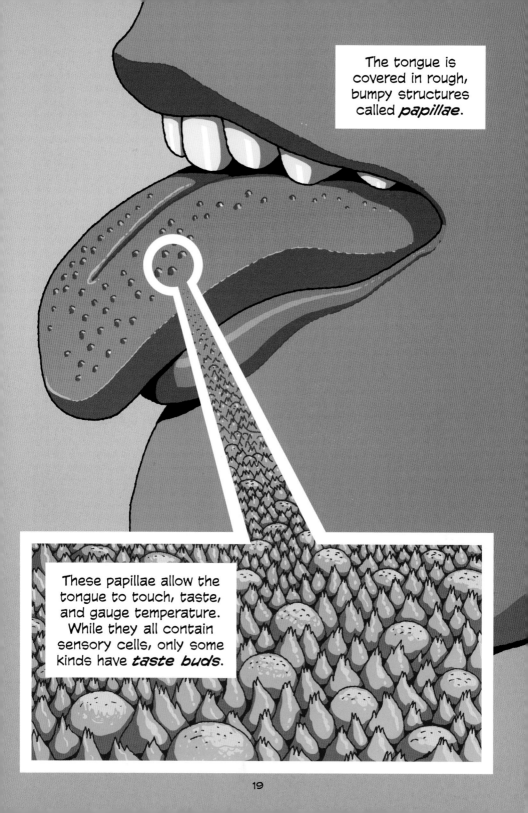

The tongue is covered in rough, bumpy structures called *papillae*.

These papillae allow the tongue to touch, taste, and gauge temperature. While they all contain sensory cells, only some kinds have *taste buds*.

Food is softened in your mouth, turning some of it into a solution. The solution flows into trenches between the papillae, reaching the opening of the taste buds.

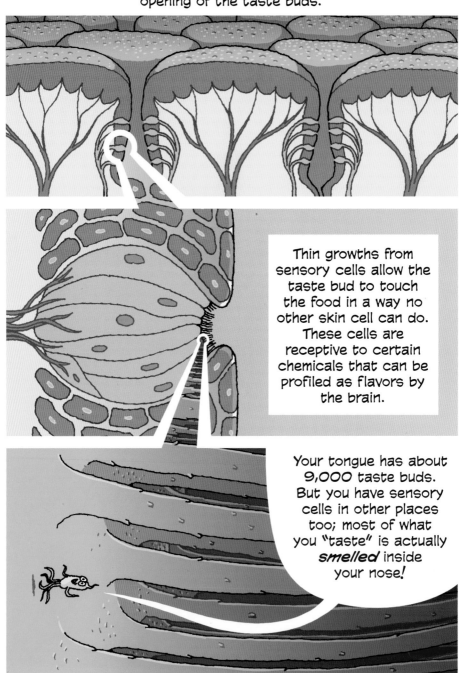

Thin growths from sensory cells allow the taste bud to touch the food in a way no other skin cell can do. These cells are receptive to certain chemicals that can be profiled as flavors by the brain.

Your tongue has about 9,000 taste buds. But you have sensory cells in other places too; most of what you "taste" is actually *smelled* inside your nose!

There are five types of taste cells, each receptive to a different type of chemical. A taste cell's neuron tells the brain which specific flavor is in the mouth.

Sweet.
Lots of sugar!

Sour.
A high level of acidity!

Salty.
A high level of, well, salt!

Umami.
Good for making proteins!

Bitter.
Potentially poisonous. But sometimes tasty!

Taste evolved to identify nutrients and warn against harmful toxins. That's why ripe fruit tastes good and poisonous plants taste bad. The more scarce and valuable a nutrient is, the more tasty it is! As hunter-gatherers, humans looked for energy-dense food and sought fats, sugars, and salt.

That's why today you desire those flavors so much, even though they are no longer scarce.

In order for solid food to be turned into a taste-able solution,
it must encounter the sensational secretion known as...

Your mouth produces
1–1.5 liters (2–3 pints)
of saliva every day! Saliva is
99% water, but also contains
antibodies, enzymes, mucus,
and other compounds. Each
person's saliva is unique!

Most of that saliva is provided on demand by
papillae located inside of your cheeks near your ear.
This saliva comes from the *parotid glands*, which are
the largest of the three major salivary glands.

You start
chewin' and I'll
start spewin'!

Below the parotid glands and beneath the floor of the mouth are glands that provide a constant supply of saliva. This saliva can't stop and won't stop.

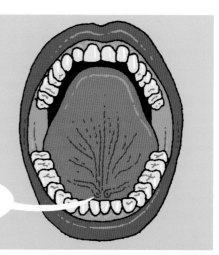

It's our party!

To produce saliva, your salivary glands pull water from your blood, then add minerals and enzymes like—

Excuse me! This is the third time you've mentioned enzymes without explaining what they are!

Technically, it's the third time *you've* mentioned them too, since you're an exact duplicate of me.

POP

Whatevs. I'm outta here!

⋝SIGH⋜ I'm so lonely.

An *enzyme* is a special kind of protein that helps start and speed up a chemical reaction inside a cell. They are responsible for thousands of different chemical reactions. Each enzyme is highly specialized, only reacting with the substance it was made for.

On the surface of the enzyme is a special region that fits perfectly with specific molecules. It could be just a single molecule, or it may require more. The compatible molecules are called *substrates*.

The substrates bond with the enzyme. That makes the enzyme change its shape, breaking one substrate apart or smashing two together!

The substrates become products, forever changed.

Saliva contains an enzyme called *amylase*. Amylase breaks down starch, a complex carbohydrate, into simple sugars. That's why when you chew bread, rice, or potatoes, you might detect the sugar released as your saliva gets to work.

I'm actually quite sweet once you get to know me.

And some saliva carries an enzyme called *lipase* that starts breaking down triglycerides.

Saliva from the bottom of your mouth is sticky and thick because it contains proteins called *mucins,* a component of mucus. This allows it to cling to the surface of your teeth and gums.

There it delivers calcium and phosphate to repair the tooth enamel damaged by acid.

Thank you, saliva!

New bacteria are introduced into your mouth every time something goes into it, food or otherwise.

I just came from a nice bowl of rice!

The mucins help trap the bacteria. Antibacterial substances like enzymes and antibodies go after it.

Wait! It's just a little food poisoning...

Then other enzymes break down the bacterial cell walls! Those folks are a little dangerous, so I'm going to just keep moving...

In addition to fighting bacteria and cleaning your teeth, saliva helps heal wounds. It causes many viruses to be noninfective. And it's a natural painkiller— which is why a sore throat often feels better right after eating!

Saliva production goes way down while you sleep, which means the bacteria in your mouth can multiply and thrive, eating dead cells that would otherwise be washed away.

It's *our* time!

Tongue cells, freshly shed!

The gas we release overnight becomes what you call "morning breath."

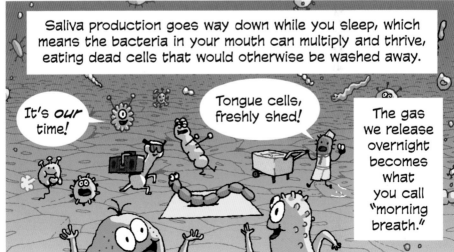

So as your jaws, teeth, and tongue masticate to change the food physically, your salivary glands provide enzymes that change the food chemically. The result is a soft, round mass called a...

BOLUS — Perfect for Swallowing!

When your tongue presses the bolus against the *soft palate* on the roof of your mouth, it triggers the swallowing reflex, a tightly coordinated chain of actions that involves more than 22 muscle groups!

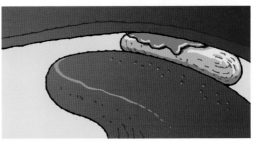

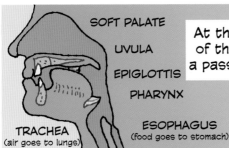

SOFT PALATE

UVULA

EPIGLOTTIS

PHARYNX

ESOPHAGUS
(food goes to stomach)

TRACHEA
(air goes to lungs)

At the back of the mouth is a part of the throat called the *pharynx*, a passageway for both food and air.

The *uvula* (that projection dangling from your palate) extends to the back of your throat to prevent food from passing through the wrong side of the pharynx. If the uvula can't seal the opening properly, food can wind up in your nose.

When the bolus is pushed into the pharynx, the tongue blocks the way back into the mouth. Now every precaution is made to try to stop you from inhaling your food.

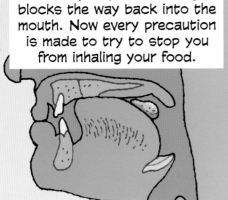

ACHOO!

Next, your vocal cords close as your voice box (the larynx) is pulled up.

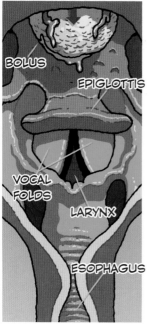

BOLUS

EPIGLOTTIS

VOCAL FOLDS

LARYNX

ESOPHAGUS

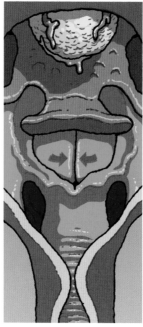

Attached to the entrance of the larynx is the *epiglottis*, which rises and flips a lid over it, closing off your windpipe. If food gets in your windpipe, you'll choke.

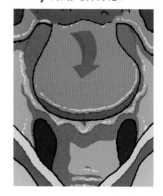

At the back of the pharynx is a sphincter that opens up and lets the bolus through.

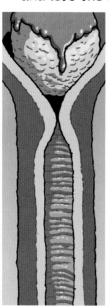

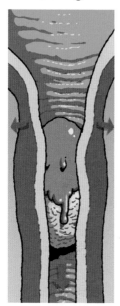

And it's gone! That's everything you need to know about digestion.

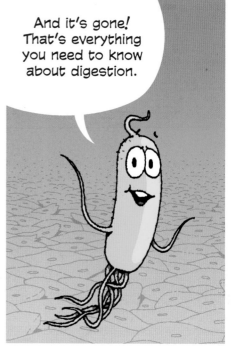

I hope you've enjoyed this tour of the oral cavity and gained a new appreciation for the organs, tissues, and glands that make eating possible. If you could please take just a few minutes to fill out this comment card, I would value your feedback!

Oral cavity Tour

please rate the Following on a Scale of 1-5

cleanliness of mouth:
Professionalism of the guide:
Metaphor Quality:
Joke balance:

is there anything we could have done To Make your experience better?

That was a great introduction.

More than an introduction, really; there isn't anything else to cover.

But this place is just the *start* of the digestive system.

The DIGESTIVE SYSTEM! It's a long, muscular tube called the *gastrointestinal tract* with many more organs than just those in the mouth. Like the esophagus! The stomach! The intestines! The pancreas, liver, and gallbladder!

Digestive... *system?*

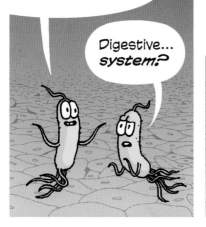

How do you not know this? The whole GI tract is about 5 meters, or 16 feet, long, and the real party is at the end of the tube!

What's at the end of the tube?

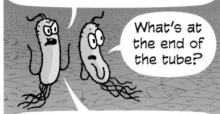

The Big City! It's where most of us bacteria live. Why are you hanging around this Podunk place? Get yourself swallowed!

You're not interested in all that, are you? Let's see... have I talked about the *mucosa* yet? The oral mucosa is a thick lining of cells and tissue that covers the inside of the mouth! It's where I live along with a lot of other bacteria and, *uh*...

Let's just get away from that pharynx, right?

Yeah, so, some people chew more on the right side, some on the left...

Some chew fast and short, some—

BOLUS!!!

I'm being swallowed! Somebody help! *HELP!!!*

Calm down, kid. You're just going to the gullet.

What's a gullet?! Oh no!

I'll help you through! The name's Spiro.

You have experience with gullets?

Ha! Would I be here if I didn't?

Um...

Well, maybe not *direct* experience. But I've spent my life studying our collective knowledge in anticipation of this very day! I've even drawn up a map...

I love maps!

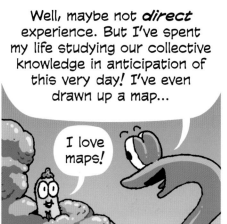

Then you shall join me in my quest for fortune and glory! That sphincter just opened, and we're now heading directly into...

the *ESOPHAGUS!*

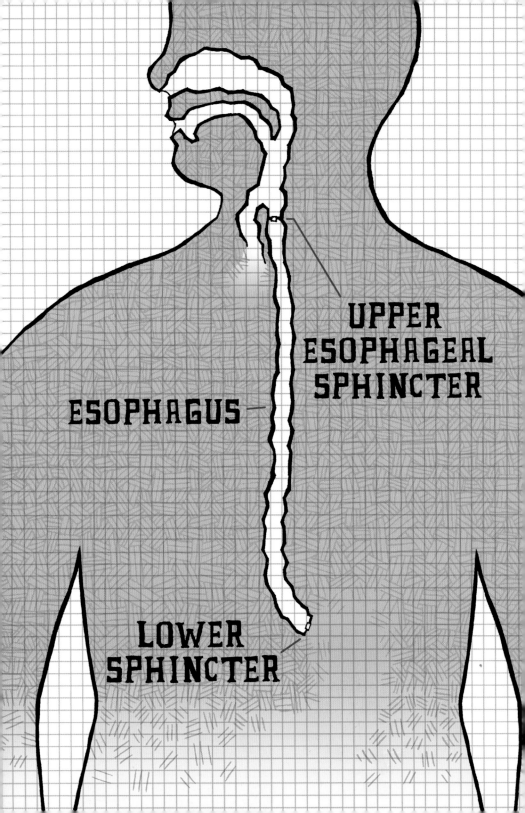

UPPER
ESOPHAGEAL
SPHINCTER

ESOPHAGUS

LOWER
SPHINCTER

Connecting the mouth to the stomach, the esophagus is 2 cm (1 inch) wide and 25 cm (10 inches) long. The top third is made of striated muscle just like the tongue.

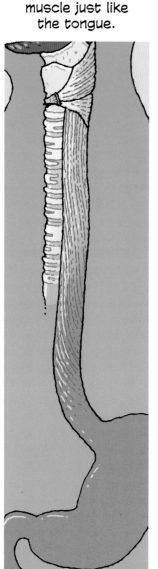

The lower part is made of smooth muscle, which you can't control consciously.

When the bolus arrives, the smooth muscles contract behind it.

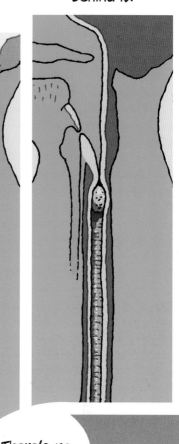

To propel the bolus to the stomach...

...the muscles squeeze to push it down and widen to let it through...

...contracting and relaxing in a wave. This process is called *peristalsis*.

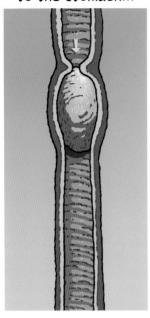

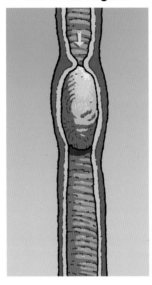

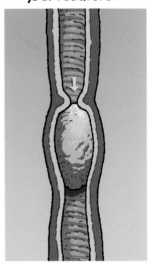

If the bolus gets stuck, signals are sent to the brain to send another wave.

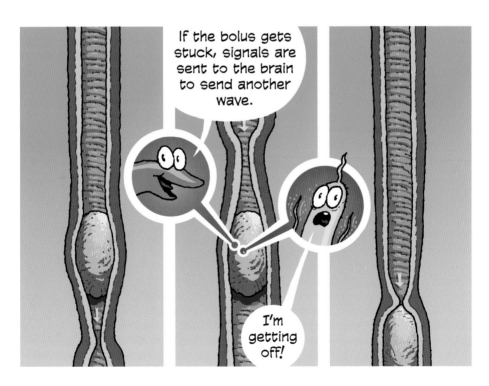

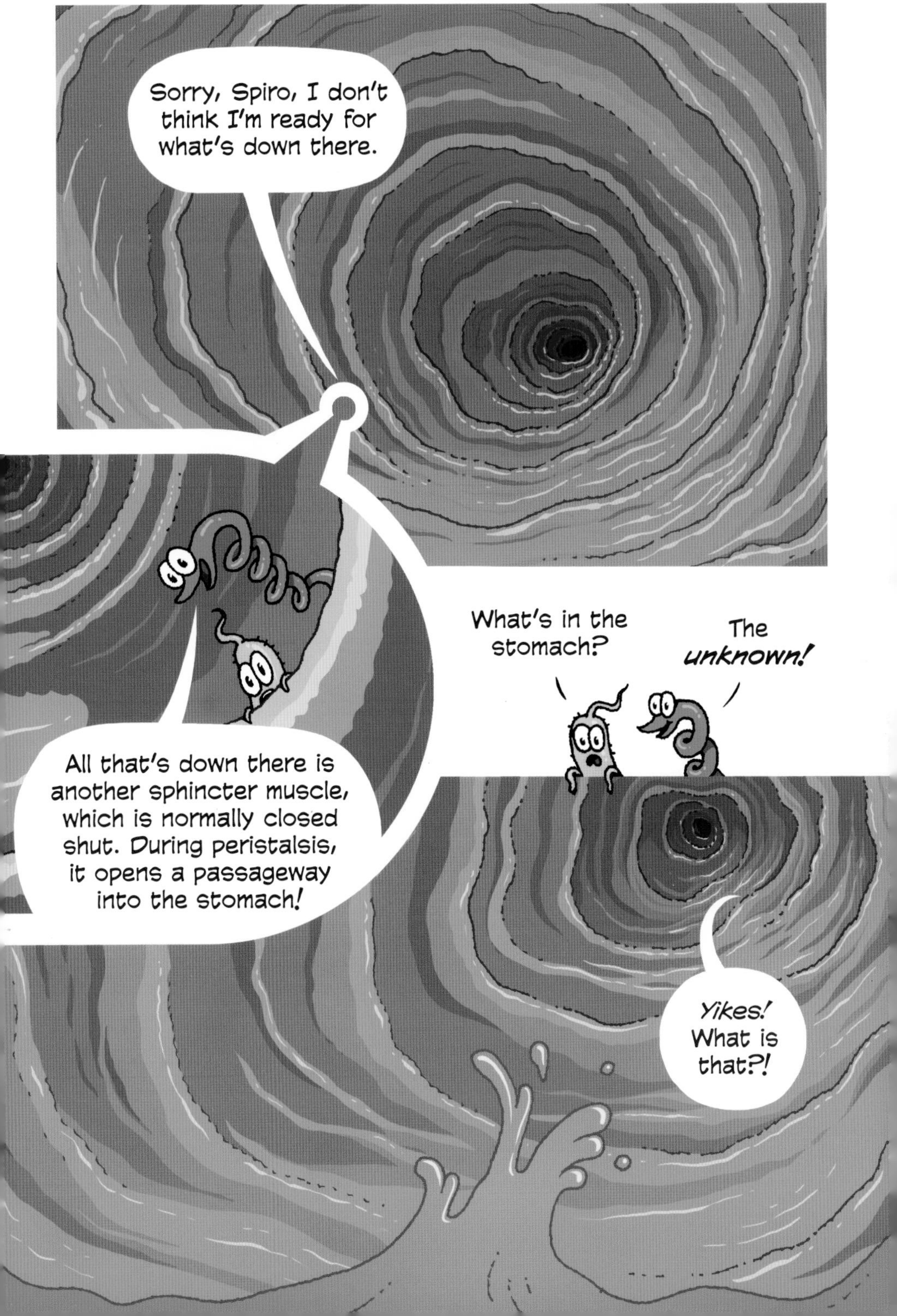
Sorry, Spiro, I don't think I'm ready for what's down there.
All that's down there is another sphincter muscle, which is normally closed shut. During peristalsis, it opens a passageway into the stomach!
What's in the stomach?
The *unknown!*
Yikes! What is that?!

Oh, that's just *acid reflux*. If the sphincter down there doesn't have a good seal, liquid content from the stomach can spew up. That liquid contains acid, which doesn't belong in the esophagus— it can cause pain known as heartburn!

Acid?!

What is this place? Are you sure it's safe to venture into the stomach?

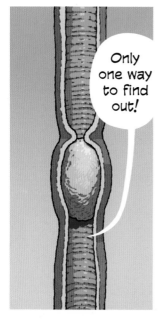

Only one way to find out!

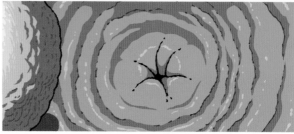

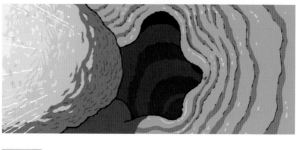

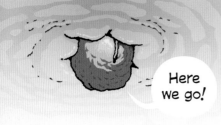

Here we go!

AAAIIIIEEEE!!

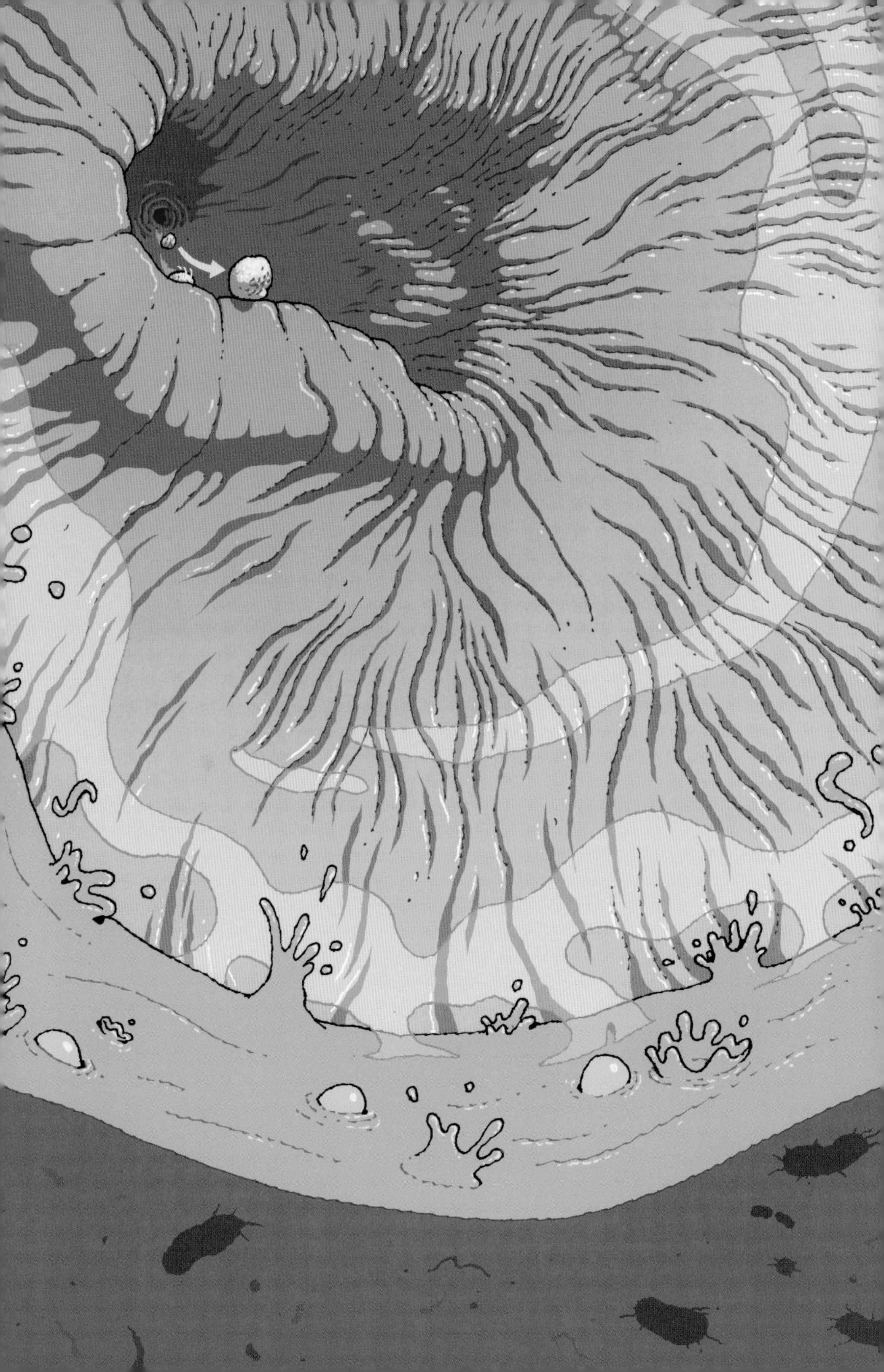

RUMMBLE
sptoooo...
sploip
SSSSSSSS
Excuse me! We're stuck!

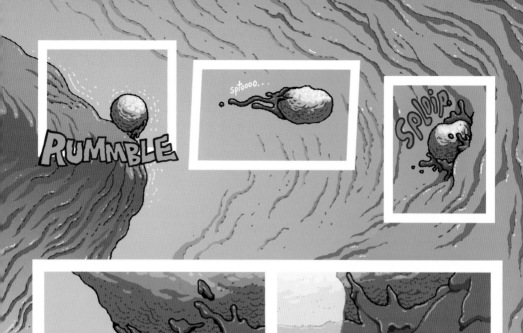

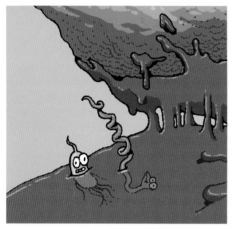

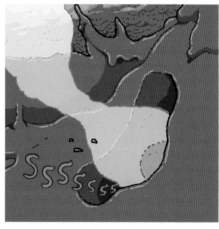

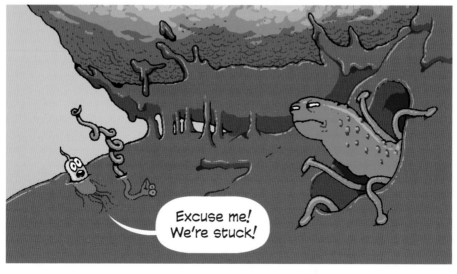

I'm *H. pylori*, navigator of the **gastric mucosa**.

Do you know anything about the stomach?

Who wants to know?

I'm E, and this is Spiro. I'm on my way to the Big City, where I can reunite with my kin. I'm also on a mission to teach our reader about digestion.

And I'm searching for adventure!

Adventure, *eh?*

Well, you've found yourselves in the most treacherous organ of the whole gastrointestinal tract. Few survive the **gastric acid** produced from the stomach lining, as it will tear apart the bonds of the weak and scatter their remains in the deep, murky depths below.

You want to learn about the stomach? 15–25 cm (6–10 inches) long with a capacity to expand *fifty to eighty times its original volume!* Aye, I shall describe its wicked ways of turning a bolus into an unrecognizable soup called **chyme**. But only the brave and the foolhardy look for adventure here.

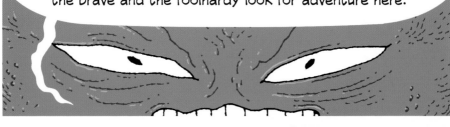

This bacterium is intense.

Prepare to meet your fate in... the **STOMACH!**

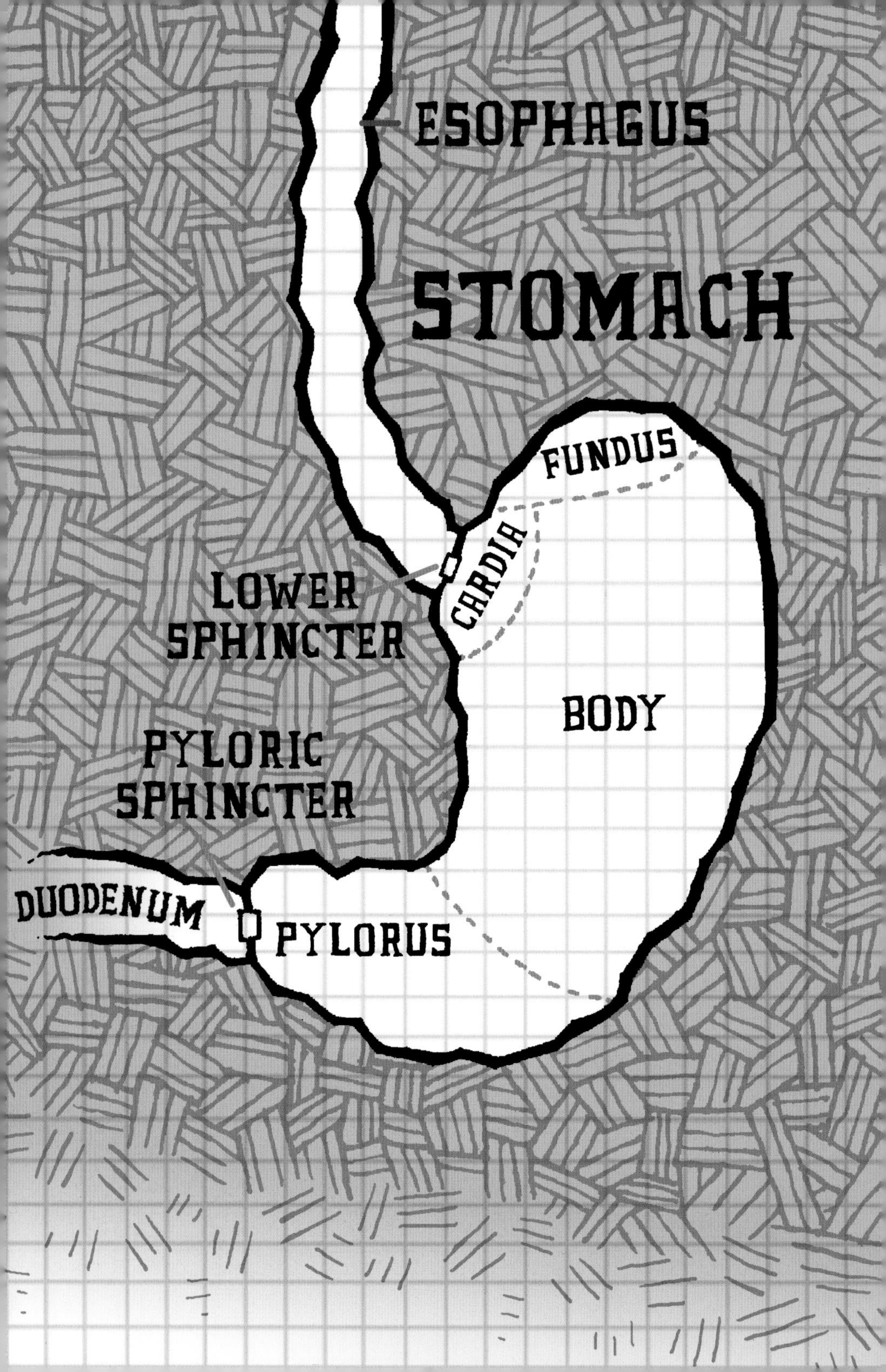

ESOPHAGUS
STOMACH
FUNDUS
CARDIA
LOWER SPHINCTER
BODY
PYLORIC SPHINCTER
DUODENUM
PYLORUS

The stomach is an elastic organ that stores food and breaks it down. Layers of muscle allow it to stretch and constrict as needed, depending on how much food is in there.

When the stomach is empty, the walls fold into large ridges called *rugae*. These folds expand outward as food arrives, providing more surface area.

The average stomach is about the size of a fist and, when expanded, can hold 2–4 liters (1/2–1 gallon). How much it holds depends on the individual body, although thin people don't necessarily have smaller stomachs than obese people.

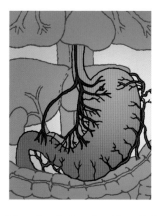
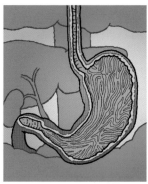
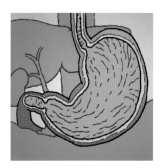

My dominion is the thick gastric mucosa, or *mucous membrane*. Mucus lines the walls of the entire GI tract. But unlike other bacteria stuck living on its surface, I have an enzyme that allows me to liquefy it and explore its inlets and outlets.

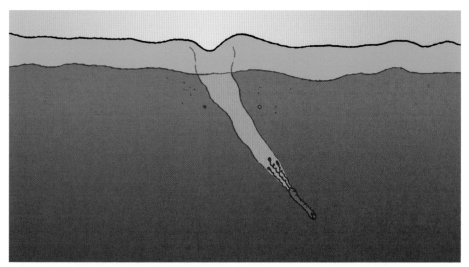

Although often associated with the nose, most mucus is produced by cells in the gut. Mucus is a gooey mix of proteins, enzymes, and salts that covers the inside of the entire GI tract, helping to move everything along.

Mucus contains mucins (just like in saliva).
So it's slippery but also thick and sticky, trapping and killing unwanted substances and preventing a lot of stuff (including us) from getting inside the rest of the body.

I have a bad feeling about this...

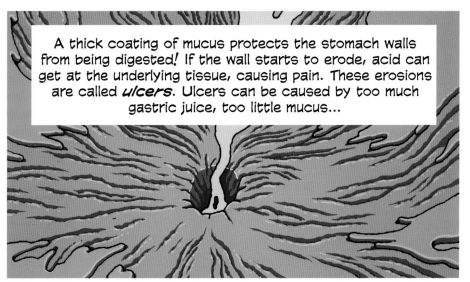

A thick coating of mucus protects the stomach walls from being digested! If the wall starts to erode, acid can get at the underlying tissue, causing pain. These erosions are called *ulcers*. Ulcers can be caused by too much gastric juice, too little mucus...

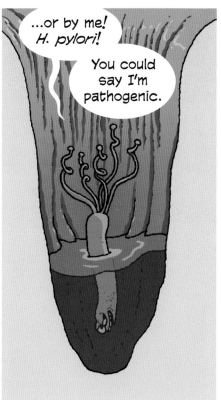

...or by me! *H. pylori!*

You could say I'm pathogenic.

The gastric mucosa features deep depressions called **gastric pits**, which contain cells that secrete substances produced by gastric glands.

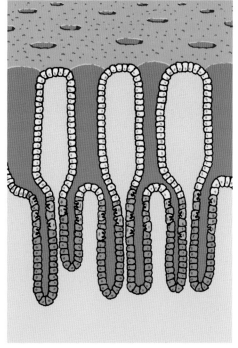

When the stomach is empty, cells in the pits release a hormone called *ghrelin*, which makes you feel hungry. If food is anticipated, other cells secrete the hormone *gastrin*, which tells everyone to get to work.

Just caught a whiff of gastrin! I'm releasing some acid to start breaking apart the food. You want in on this, Chief?

You read my mind as always! With your acid, I can make our favorite enzyme, *pepsin!* Let's take down some proteins!

Ech, this is going to get messy. I better provide more mucins to protect these walls from your juice.

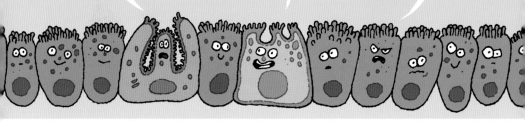

You do not want to hear the tales; I have seen this ruthless acid tear apart stronger bacteria than you. It is diabolical.

We are not afraid!

Why did I ever trust you?

Listen closely.

You are in the *fundus*.

Food can be stored here until there is room below in the body of the stomach. While in the fundus, salivary enzymes like amylase and lipase continue their work breaking apart starches and fats.

The peristalsis you saw in the esophagus occurs in the stomach as well, with the addition of a third layer of muscle that runs in a spiral. This innermost layer allows for a twisting motion when all three contract.

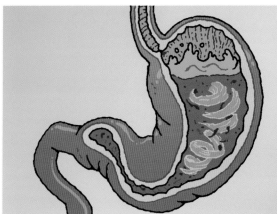

This causes a churning activity that grinds the bolus into pieces as the acid and pepsin break chemical bonds...

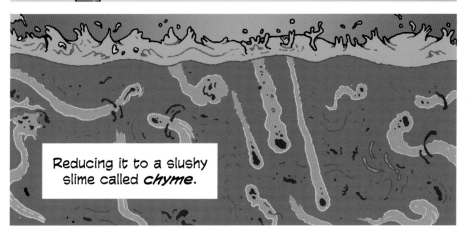

Reducing it to a slushy slime called *chyme*.

This typically goes on for about four hours.

Who's got that kind of time? I'm hitching on to the next bolus that comes by.

What? No!

Fortune and glory await!

Four hours? It took like ten seconds to get down here.

Could be two hours, could be six. It all depends on the discretion of the *pyloric sphincter*.

Another sphincter?

AAAUUGGHH!!!

The acid! It burns! Save yourself, friend— I'm done for!

AAAIIIEEEeeee....

Spiro is... fine, right?

Probably?

Of course not! Have you listened to nothing I've said? Gastric acid is one of the most corrosive substances you can find!

Anyway, where was I...

In the furthest reaches of the stomach, you shall find the *pylorus*. The Greeks call it the Gatekeeper.

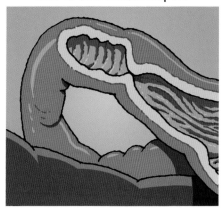

Only the meekest scraps of food enter the pyloric canal, at the end of which awaits the *pyloric sphincter*, which requires all food desiring entry to pass a test.

Are you the Keymaster?

It only allows a small amount of chyme through every few minutes. The sphincter gets closed when greeted with too many proteins or lipids; they're held back for longer processing. This is why people often feel fuller longer when they eat meat and fat instead of pasta or rice.

That feeling of fullness is caused when the stomach has stretched past a comfortable point.
Then it's time to stop eating. The sphincter at the top of the stomach relaxes, venting air up through the esophagus, causing a...

The burp comes from a *gastric bubble* in the fundus, formed from swallowed air.

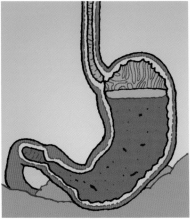

The reason the esophagus is attached to the side of the stomach instead of directly on top is to prevent air from constantly escaping.

Burp...

BURP...

burp...

BELCH!...

That makes me wonder... once in a while, blood drains out of the cheeks and a ridiculous amount of saliva fills the mouth. Soon after, muscles near the throat lift and a hot porridge gushes out!

That saliva is sent to protect your teeth from the gastric acid contained in the substance known as...

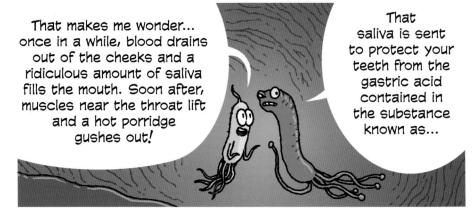

Before it is filled to bursting, the stomach will send signals to the brain that it's taking in more than it can put out. Even if it's not full, receptors in the stomach wall test the contents and send info to the brain, which tries to decide if the mouth has sent in a poison or a pathogen that the immune system can't handle.

They're ringing off the hook!

Send excess blood to the abdomen!

Mayday!

The stomach begins to move in small waves.

The pyloric sphincter opens back up to let contents back in.

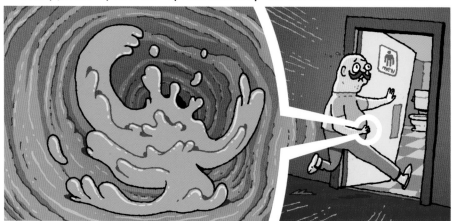

The stomach muscles squeeze as the esophagus opens its doors.

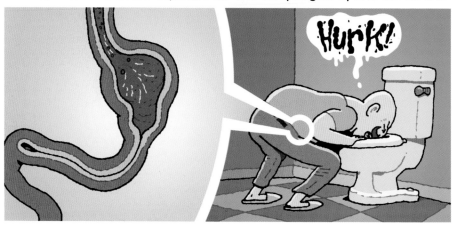

Wow! No wonder it often has some bits of food that I'd seen earlier!

Now, sudden vomiting is usually caused by a virus. With food poisoning, the brain first provides a warning through the feeling of *nausea*, a very clear signal to stop eating.

Nausea also accompanies motion sickness. The eyes report movements that contradict the balance felt by the body. The brain associates that experience with poison and acts accordingly.

So if vomit can come from the small intestine...

Bitter stuff. Yellow-green. Full of *bile*.

Does that mean the path there is through the... pyloric sphincter?

You have learned much.

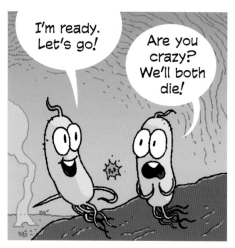

I'm ready. Let's go!

Are you crazy? We'll both die!

Who told you that? Not all bacteria are destroyed by acid. My cell has a membrane that protects me.

Come on, we're carbon copies, right? If you can't trust yourself, who can you trust?

Fear the infernal gastric juices!

My duplicate has a point...

Here goes nothing.

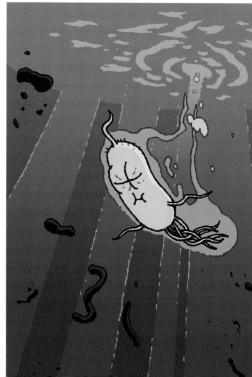

YOLO!

Gracious pyloric sphincter...

um...

I am true of heart, having traveled a great distance to pass through your—

Okay!
Not so tough.
SLAM!
Whoa...

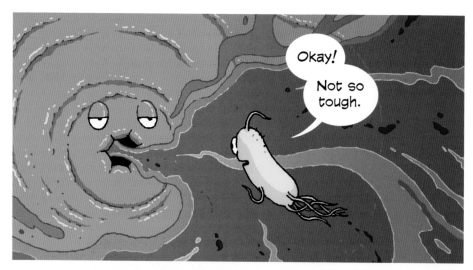

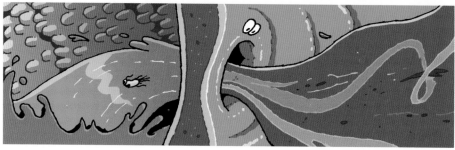

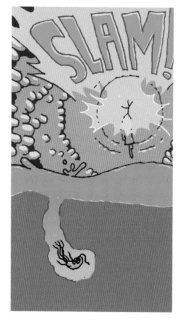

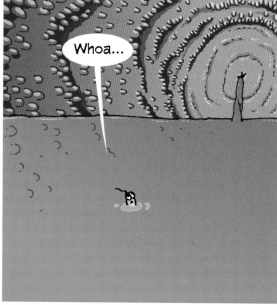

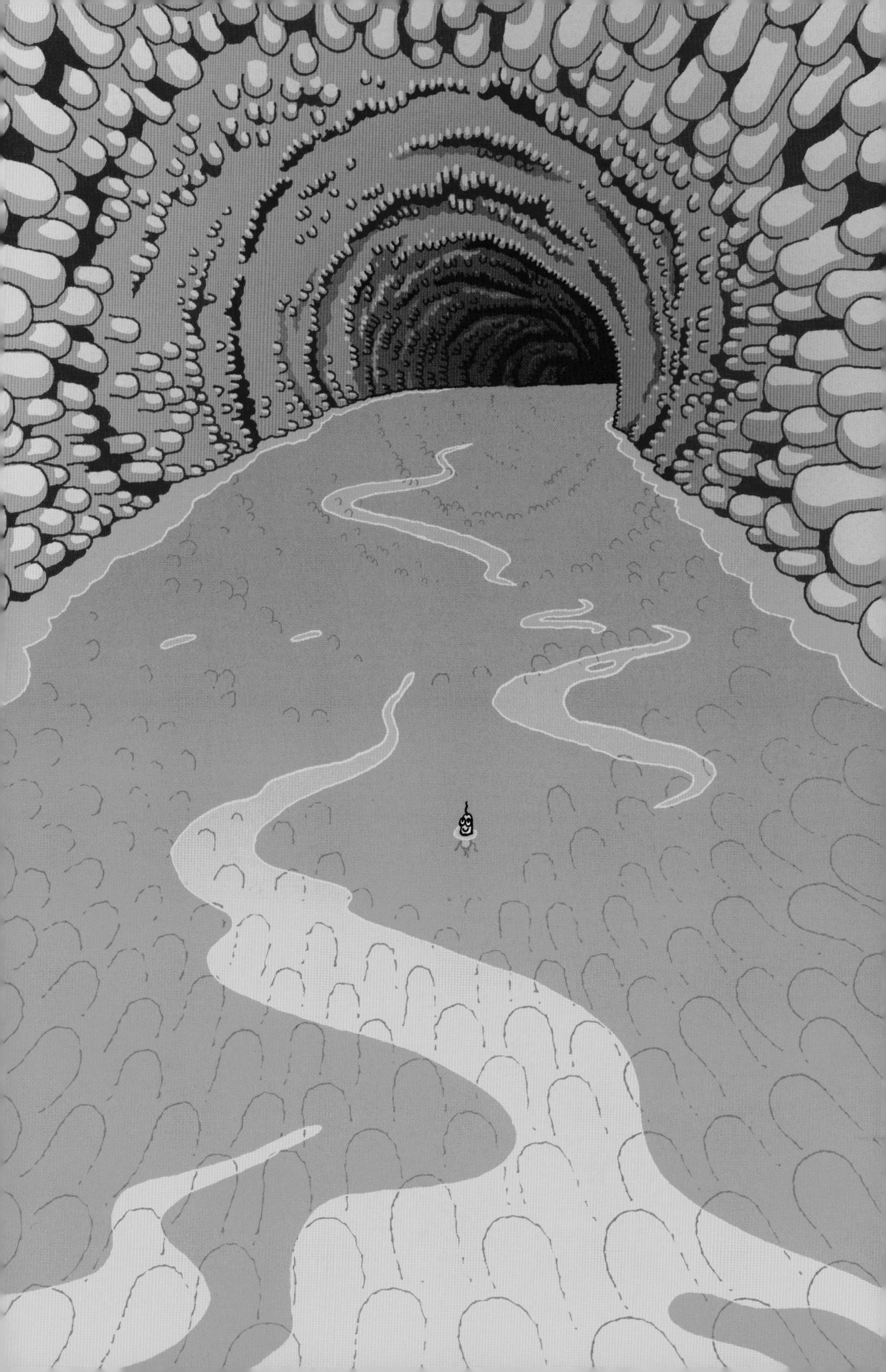

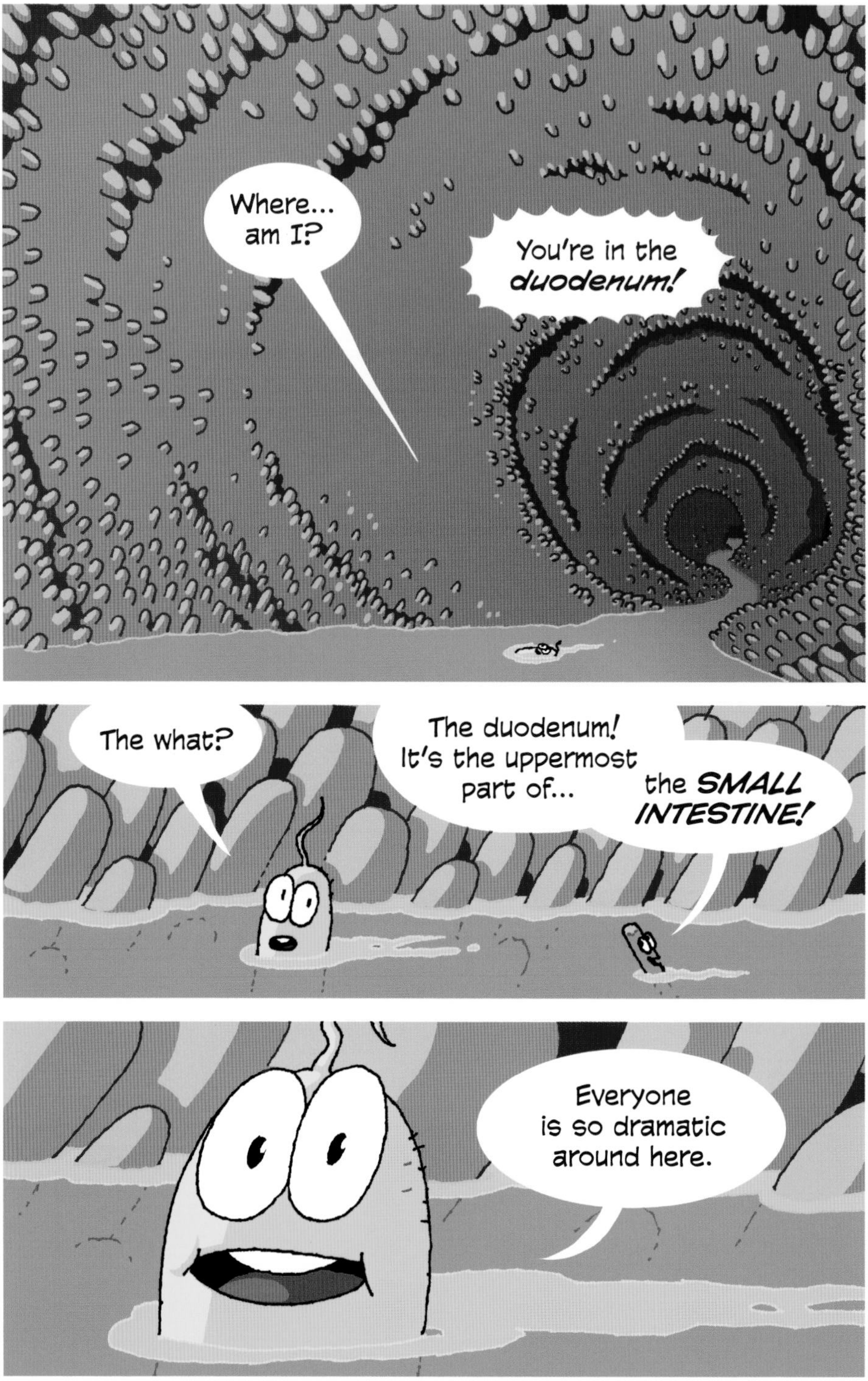
Where... am I?
You're in the *duodenum!*
The what?
The duodenum! It's the uppermost part of...
the ***SMALL INTESTINE!***
Everyone is so dramatic around here.

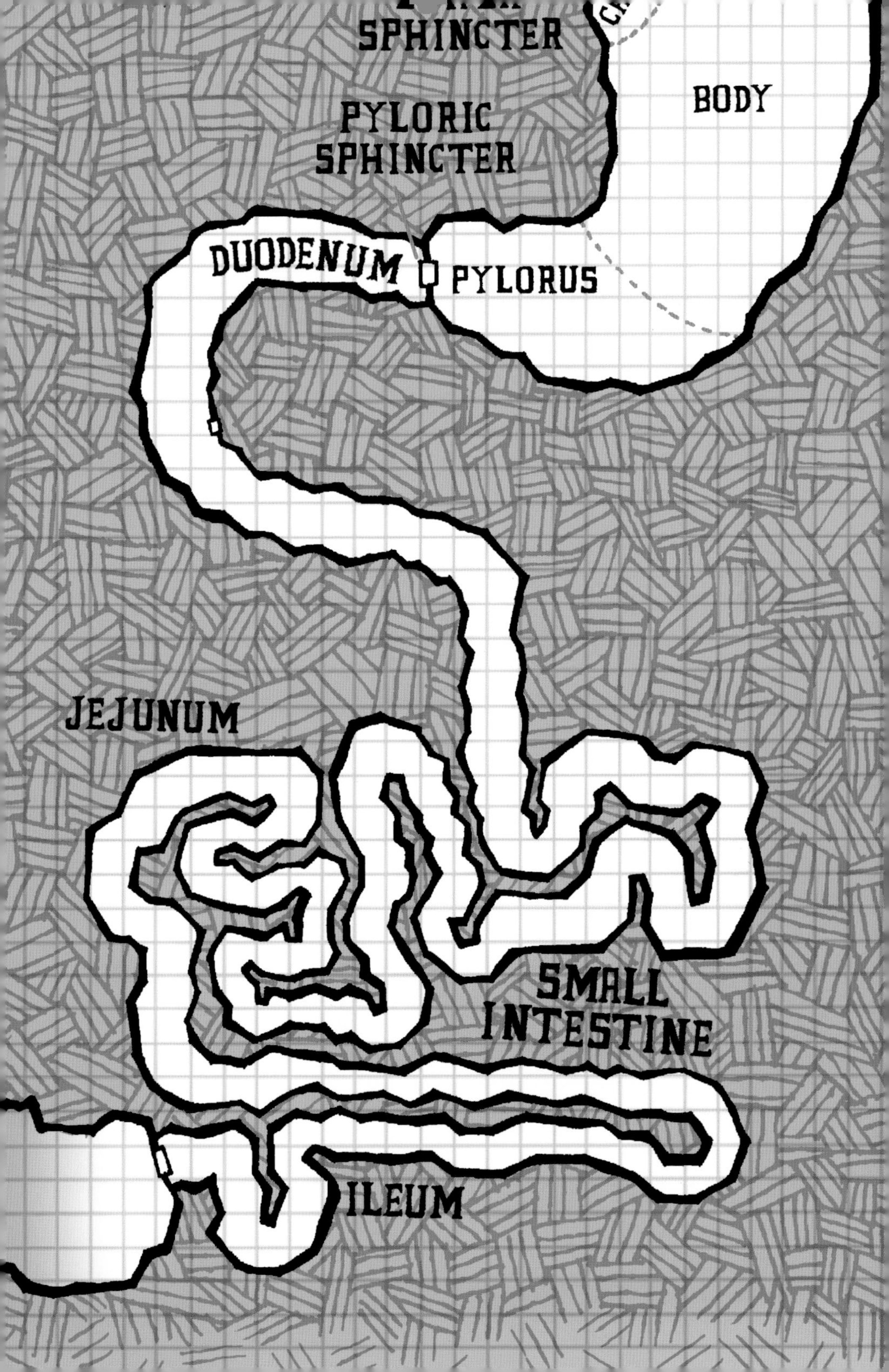
SPHINCTER
BODY
PYLORIC
SPHINCTER
DUODENUM
PYLORUS
JEJUNUM
SMALL
INTESTINE
ILEUM

The small intestine is a twisting, turning loop-the-loop that's about 2–4 meters (7–13 feet) long. It's where most of digestion and absorption takes place! By the time it reaches the end, the food has virtually disappeared.

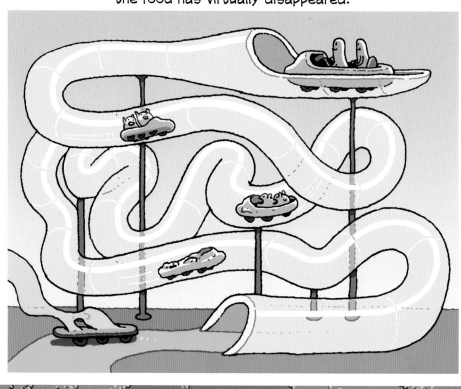

As long and winding as it is, the surface area is increased even more by folds in the mucosa.

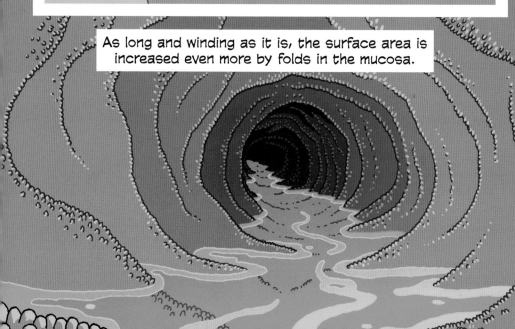

This allows for more contact with villi. The place is covered with these protrusions, about 30 per square millimeter. Nutrient absorption happens through special cells on the surface of the villi.

Each cell contains even tinier protrusions that contribute to a surface area called a *brush border*.

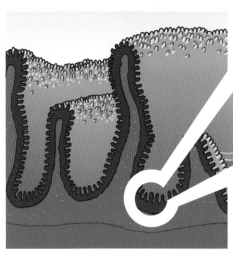

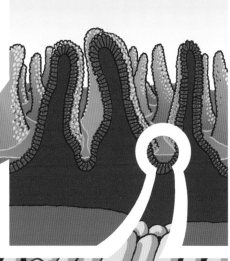

This all increases the surface area 600 times. If you stretched the brush border out flat, it could cover a tennis court!

I appreciate how readily you provided all this information!

It's what I do.

Really? I give tours too!

Yeah?

My friends all hang out in the jejunum, stopping the growth of bad bacteria while making an enzyme called *lactase* that breaks down milk sugars.

We're called *Lactobacillus*.

But I honestly find this job to be more rewarding.

In the small intestine, chyme is urged forward through peristalsis, just like in the esophagus! Muscles rhythmically squeeze and release, helped along by all the villi moving up and down.

UNCE UNCE UNCE UNCE UNCE UNCE UN

Fiber is helpful here because it can't get digested. Instead, it encourages peristalsis by pushing up against the intestinal wall...

...which pushes right back.

Keep moving, creep!

Don't stop the party!

I talk about fiber too when I cover nutrition! So how does the small intestine digest and absorb all the nutrients?

Digestion mostly happens in the duodenum using enzymes from the brush borders and other organs. Then the resulting nutrients can be absorbed in the jejunum and ileum.

If you know about nutrition, you must be familiar with macronutrients?

Yes, carbohydrates, proteins, and lipids!

Right, so...

Complex carbohydrates are held together by bonds that can be broken by brush border enzymes like amylase (also found in saliva) and lactase (like my friends make).

Proteins can be broken apart by *proteases* like pepsin (from the stomach).

And lipase (similar to the kind found in saliva) breaks triglycerides apart into separate *fatty acid chains*. But first, it needs—

Whoa, what is that?!

Oh, exactly! We've reached the middle of the duodenum.

This bile is unleashed by a muscular valve above—another sphincter!

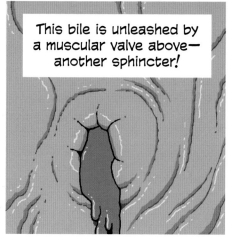

Bile is a fluid made of water, pigment, fats, and *bile salts*, which have two sides.

> I just love water. It is so attractive to my lifestyle.

HYDROPHOBIC

HYDROPHILIC

> I hate getting wet! Oh dear!

They gather around fats, the hydrophobic side facing toward the fat particles, while the hydrophilic side enjoys the water flowing by.

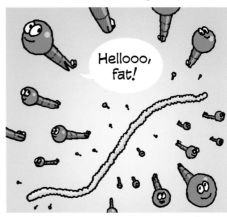

> Hellooo, fat!

These groups split the fat into chunks, the salts forming the surface of a sphere.

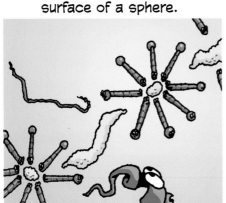

Lipase can now get at the triglycerides through gaps in between the bile salts.

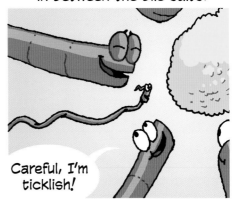

> Careful, I'm ticklish!

> Where does all this bile come from?

> I'm so glad you asked! We didn't even rehearse this.

> I'm a professional.

It is stored in one of the accessory organs of the digestive system...

the *GALLBLADDER!*

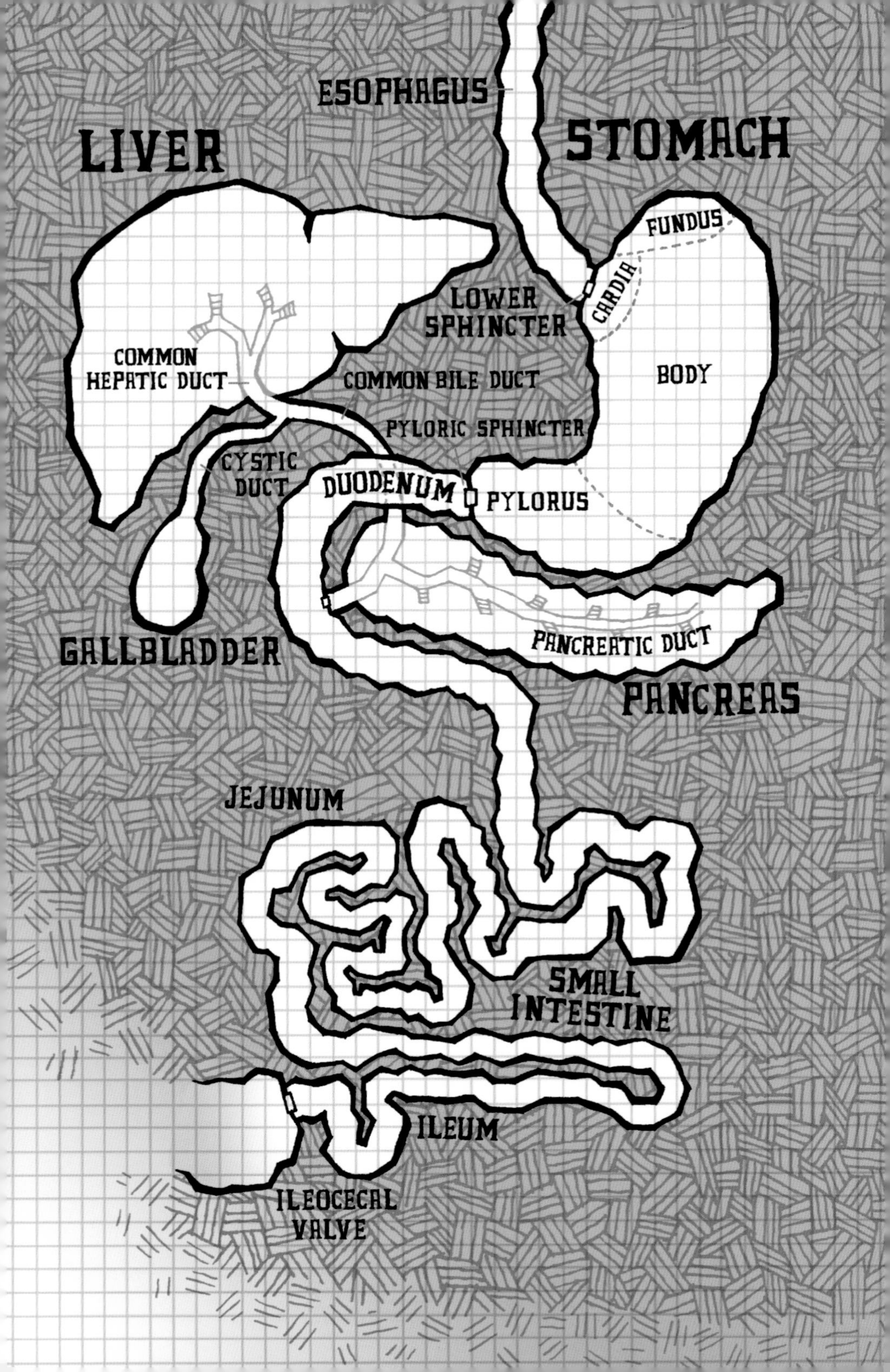
ESOPHAGUS
LIVER
STOMACH
FUNDUS
CARDIA
LOWER SPHINCTER
COMMON HEPATIC DUCT
COMMON BILE DUCT
BODY
PYLORIC SPHINCTER
CYSTIC DUCT
DUODENUM
PYLORUS
GALLBLADDER
PANCREATIC DUCT
PANCREAS
JEJUNUM
SMALL INTESTINE
ILEUM
ILEOCECAL VALVE

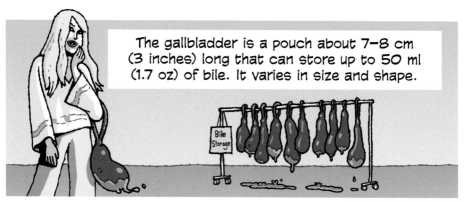

The gallbladder is a pouch about 7–8 cm (3 inches) long that can store up to 50 ml (1.7 oz) of bile. It varies in size and shape.

When cells in the walls of the duodenum detect a lot of fatty foods, they send out a hormone called...

Also known as CCK, the hormone travels to a place called the *biliary tree*.

Cholecystokinin!

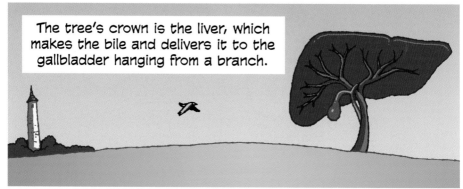

The tree's crown is the liver, which makes the bile and delivers it to the gallbladder hanging from a branch.

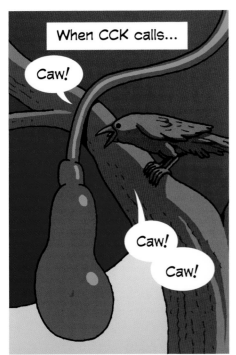

When CCK calls...

Caw!

Caw!

Caw!

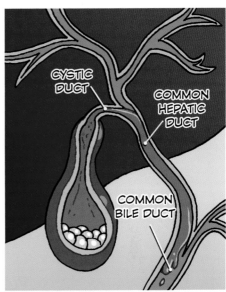

The muscles of the gallbladder contract, sending bile from the liver into one of the branches...

CYSTIC DUCT

COMMON HEPATIC DUCT

COMMON BILE DUCT

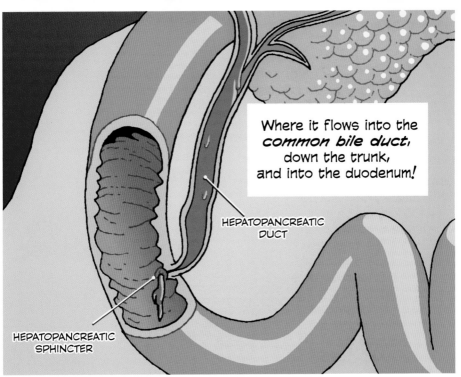

Where it flows into the *common bile duct*, down the trunk, and into the duodenum!

HEPATOPANCREATIC DUCT

HEPATOPANCREATIC SPHINCTER

When stored in the gallbladder, some water is removed from bile, concentrating the other ingredients. If cholesterol or pigment becomes too concentrated, hard crystals called *gallstones* are formed. Most gallstones are harmless, either staying in the gallbladder or just passing through the system.

Occasionally, one will get stuck in a duct. This is painful. It could pass on its own in several hours. For recurring problems, the gallbladder may be surgically removed.

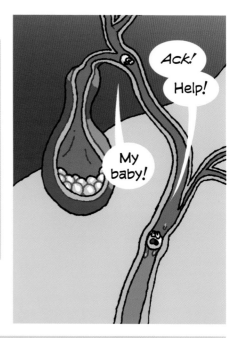

> I'm tired of just waiting around in this boring old sack! I'm leaving with the next contraction.

> No, honey, it's too dangerous!

> *Ack!*
>
> Help!

> My baby!

CCK travels with another hormone, *secretin,* in response to a meal. Their mission is to stimulate the release of a second fluid. Unlike bile, this fluid is thick and clear! It carries *bicarbonate* to neutralize the stomach acid, and also digestive enzymes produced by... the *PANCREAS!*

Unlike the other digestive organs, the pancreas is not enclosed by any layers of muscle fiber and mucosa. Instead, it is covered by thin connective tissue that separates clusters of cells.

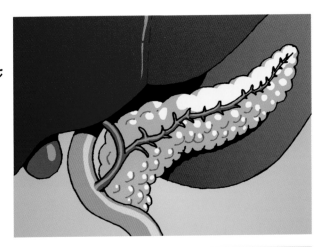

The pancreas keeps down two full-time jobs!

Show-off.

JOB #1: Its *endocrine* role involves bundles of cells called *pancreatic islets*, which release hormones that help build up and store carbohydrates and other macronutrients.

JOB #2: Its *exocrine* role directly relates to digestion, so let's take a closer look. Surrounding the islets are loose clusters of cells called *acini*.

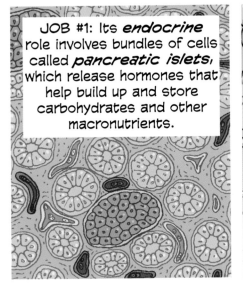

Cells in each cluster secrete enzymes that gather in the ducts.

While in the pancreas, all the enzymes are inactive. Here they're known as *zymogens*.

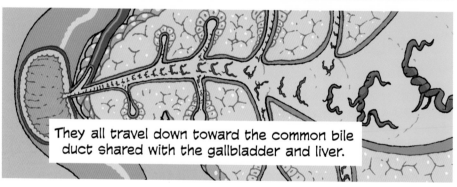

They all travel down toward the common bile duct shared with the gallbladder and liver.

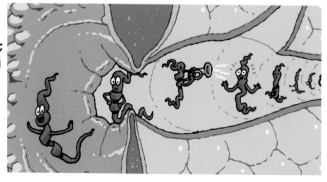

When they are released into the duodenum, they get woken up, changing into fully realized enzymes. Some of the enzymes, once activated, wake up the rest of the zymogens!

Pancreatic juice is made of powerful stuff!
Here are some of the ingredients.

Bicarbonate neutralizes the gastric acid in chyme, making it safe to flow along the intestinal walls.

Proteases break down proteins.

Amylase breaks down carbohydrates.

Lipase breaks down lipids.

Amylase, lipase, and protease are also the main enzymes used in laundry and dish detergents! They digest the food and grease that are left behind on clothes and plates.

If the pancreatic duct is blocked, the pancreas swells up— a serious condition called *pancreatitis*. The enzymes wake up, unable to leave.

In chronic pancreatitis, the enzymes start to digest the pancreas itself, eating their way out.

They can even get loose, damaging the surrounding organs and blood vessels. Pancreatitis is treated with a lot of fluids to wash out the enzymes.

There are several potential causes, including excessive alcohol use. A stuck gallstone could also block the pancreatic duct.

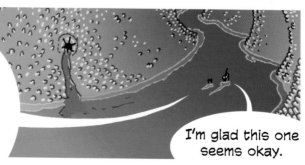

Looks like we are leaving the duodenum and about to enter the *jejunum,*

the middle part of the small intestine.

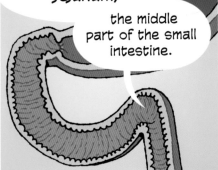

See the villi getting longer? While a lot of the breakdown work is done up in the duodenum, most of the nutrition absorption happens here in the jejunum!

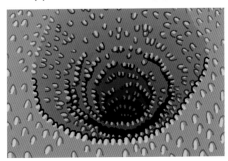

Now that proteins, complex carbohydrates, and most triglycerides have been broken down, cells can absorb amino acids, simple sugars, and fatty acids.

All nutrients are either water soluble or fat soluble. Amino acids and sugars are water soluble, so they can be sent directly through the blood. Lipids are sent through *lymph*.

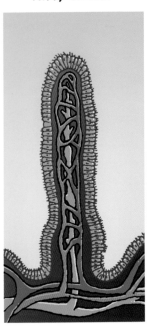

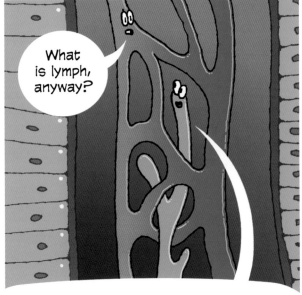

What is lymph, anyway?

Lymph is the clear fluid from blood that contains things like lipids wrapped in proteins and salts, along with junk leaked from cells all over the body. It's where the immune system screens fluids for signs of infections.

Lymph delivers bits of proteins to lymph nodes for inspection. Sometimes, even though a protein may be safe, the body confuses it as a sign that a major infection is underway.

Papers, please.

That's what the last peanut said!

Uh...I...lost my identification?

There's been a mistake!

I'm harmless!!!

Run it up the chain. Tell them not to let any more peanuts past the oral cavity!

Blood vessels of the small intestine all eventually converge (along with other vessels), transporting whatever has been absorbed to a blood vessel behind the neck of the pancreas called the *hepatic portal vein*.

At the other end of this vein, the packages will undergo a robust screening process in the largest and heaviest organ inside the body...

the *LIVER!*

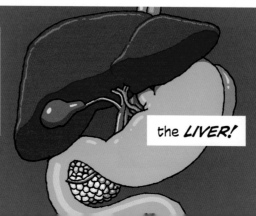

The liver is a 1.5 kg (3.5 lb) powerhouse about 15 cm (5 inches) wide. It performs about 500 separate functions!

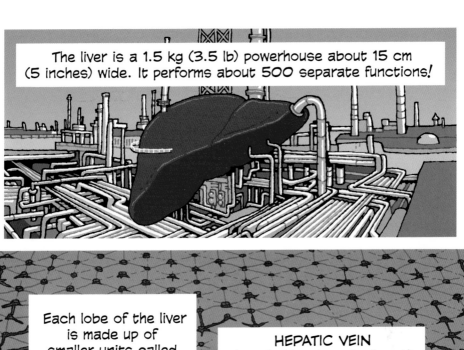

Each lobe of the liver is made up of smaller units called *hepatic lobules,* which is where all the action happens. The lobules are surrounded by small branches to blood vessels; collectively each group of branches is called a *portal triad.*

HEPATIC VEIN
(sends blood to the heart)

COMMON HEPATIC DUCT
(sends bile to the common bile duct)

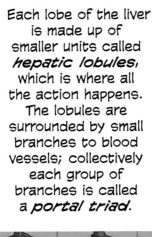

PROPER HEPATIC ARTERY
(brings in oxygen-rich blood)

PORTAL VEIN
(brings in nutrient-rich blood)

Filling each lobule are millions of liver cells that extract nutrients from the blood brought in from the portal vein. From the nutrients, they can assemble fats and molecules like complex carbohydrates, or break them down for use.

This is where energy stores are created from carbohydrates when we eat too much! They are stored as glycogen or converted into lipids.

The lipids delivered by lymph get processed and repackaged so they can be sent into the bloodstream.

Toxins are found and destroyed before the circulatory system has a chance to send them through the body.

They do this with the help of a protein called *cytochrome P450* (CYP for short). This enzyme is special because it doesn't need a specific substrate to bind with; it can react with a wide variety of substrates!

I accept molecules for who they are.

I recognize their inner beauty...

...and I bring it to the surface.

CYP can handle a lot of different substances, transforming them and separating out the toxins.

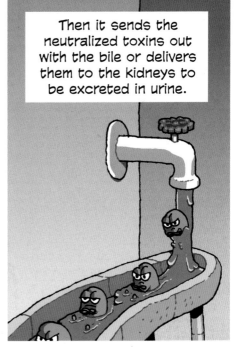

Then it sends the neutralized toxins out with the bile or delivers them to the kidneys to be excreted in urine.

In the center of the lobule, blood picks up the toxin-free nutrients, preparing them for delivery to cells all over the body. The liver supports every other organ with energy released from those nutrients!

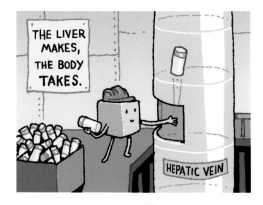

The liver is vulnerable to many diseases, like *hepatitis*, an inflammation of the tissue. The immune system sends its crew to a damaged or diseased part of the liver, which tries to remove the dead cells and damaged tissue.

If this happens regularly, scar tissue replaces the healthy liver tissue, which can block the flow of blood. This limits the liver's abilities and causes serious problems!

If any part of the liver is destroyed or removed, it can regenerate! It duplicates the remaining lobules in order to restore the function of the original liver. It needs to do this because it plays such a critical role; no one can survive long without a liver.

We've reached my colony.

We've missed you!

Welcome back!

This is a great crew! They get first dibs on all the food and push the other bacteria away.

Where have you been all this time?

You look great!

I... feel... funny...

Missed you too! I've been away so long, I almost forgot what it's like to be with my own kind.

OMG, you should totally stay!

There's always glucose in the jejunum!

What... is... happening...?

Man, it feels good to be home! Remember the time when—

Hold up, what's wrong with your friend?

It's probably all the lactic acid.

Lactic... acid...?

Yep! Another way lactobacilli inhibit the growth of other bacteria is by producing *lactic acid*, which causes protein leakage, damaging cell membranes and—

That's enough... I think I need to move on...

Oh, right! Sorry!

The last section of the small intestine is the *ileum*. Looks like the jejunum but the tube is smaller and the villi are shorter.

Good luck!

Ungh...

you've been very...generous. Thank you.

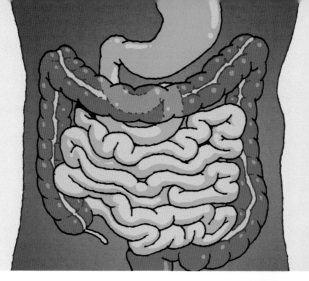

Seeing that lactobacillus so loved by its family makes me long for my own.

I wonder when we reunite if they'll remember me.

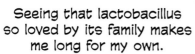

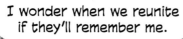

RUMBLE!

What's that?!

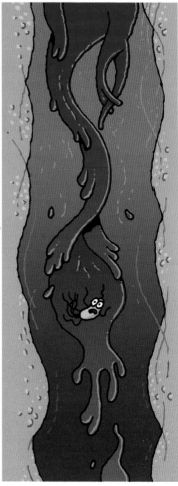

Phew! That wave was really strong!

Well, *duh*. It was part of the *migrating motor complex*.

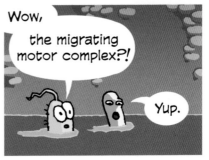

Wow, the migrating motor complex?!

Yup.

Aren't you going to teach me about it?

I mean... I wasn't planning on it.

But I guess I coul—

YEAH!

About an hour after food has been digested in the small intestine, there are still some scraps floating around in the stomach.

When are you heading out?

Eh, I'm happy bumming around here with you for a while.

No rush, just wondering—

THE PARTY IS OVER!

So the pyloric sphincter opens, forcing the leftovers through.

Hey, mind your business!

We'll leave when we're ready!

Strong peristaltic contractions begin in the stomach and move on through the small intestine, pushing all the remaining particles to the end.

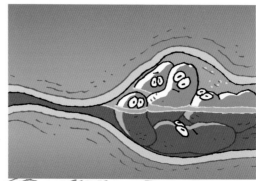

The housekeeping process is loud; people usually take a rumbling belly as a sign of hunger. But the gut isn't trying to send a message. It's simply tidying up!

RUMMBLE!!

And real hunger? When the stomach is empty and all nutrients have been absorbed, signals are sent to the brain. After several hours, the stomach will contract in a different, more painful way. Hunger is felt, not heard.

CONTRACTIONS

Hey!

Sharing that was kind of fun.

Wasn't it?

We're transmitting all of this to the reader using advanced comics technology!

It's an exciting time to be a bacterium.

I feel another wave coming...

RUMMBLE!

Hey, it's my peeps, the *Bacteroidetes!*

Welcome to the large intestine!

Ugh, why do you keep saying that?

This is the *ileum*. You're still in the small intestine.

Okay, technically.

But most of the nutrient absorption is over, so it's practically the large intestine.

It is nothing like the large intestine— look at all the villi!

You always act like you're ashamed to live here.

This place is great; it has all the fiber of the colon but none of the crowds.

There's a reason for that...

To prevent overgrowth, these suburban populations are tightly regulated. The ileum lining thickens, creating follicles that extend into the mucosa. They monitor the traffic passing through, looking for harmful molecules (*antigens*) and microorganisms (*pathogens*).

And the gateway to the large intestine is tightly controlled by a sphincter that prevents most bacteria from making its way back into the small intestine.

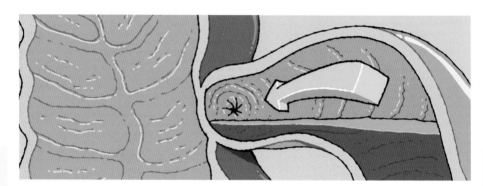

Now that the bile has done its job with the lipids, brush border membranes begin the absorption of bile salts to send back to the liver.

The large intestine moves at a more relaxed pace than the small intestine. That's partly why it's such a haven for bacteria; we're given a lot of time to chow on food that comes down the pipe.

EXIT ONLY; NO REENTRY

I've been in the small intestine for six hours now. I think it's time to move on!

I have a ton of connections in the large intestine. You should look up my cousin Bif.

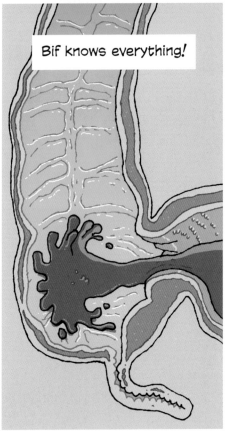

Bif knows everything!

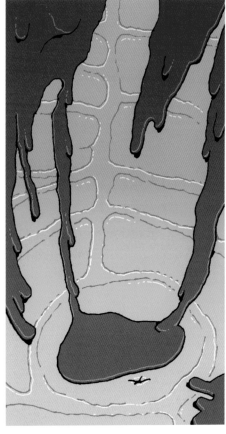

Excuse me, I'm looking for...

Outta the way!

Do any of you know a bacterium who goes by the name Bif?

Bif moved to start a colony in the transverse colon.

Where's the transverse colon?

I'm not from around here.

I can take you as far as the hepatic flexure, which will get you right into the transverse.

You can?

Headed there anyway. Lookin' to start a colony and take over this joint.

Get lost, C. *diff!*

Nobody needs your help!

I'm watching you, *Clostridium coccoides.*

Whatever!

Look, kid, you gotta be careful in the large intestine. There's over 500 different species living here...

Many are *symbiotic* and have a relationship with the other species, but some are out only for themselves.

Hand over that map.

I can give you the lowdown on...

the LARGE INTESTINE!

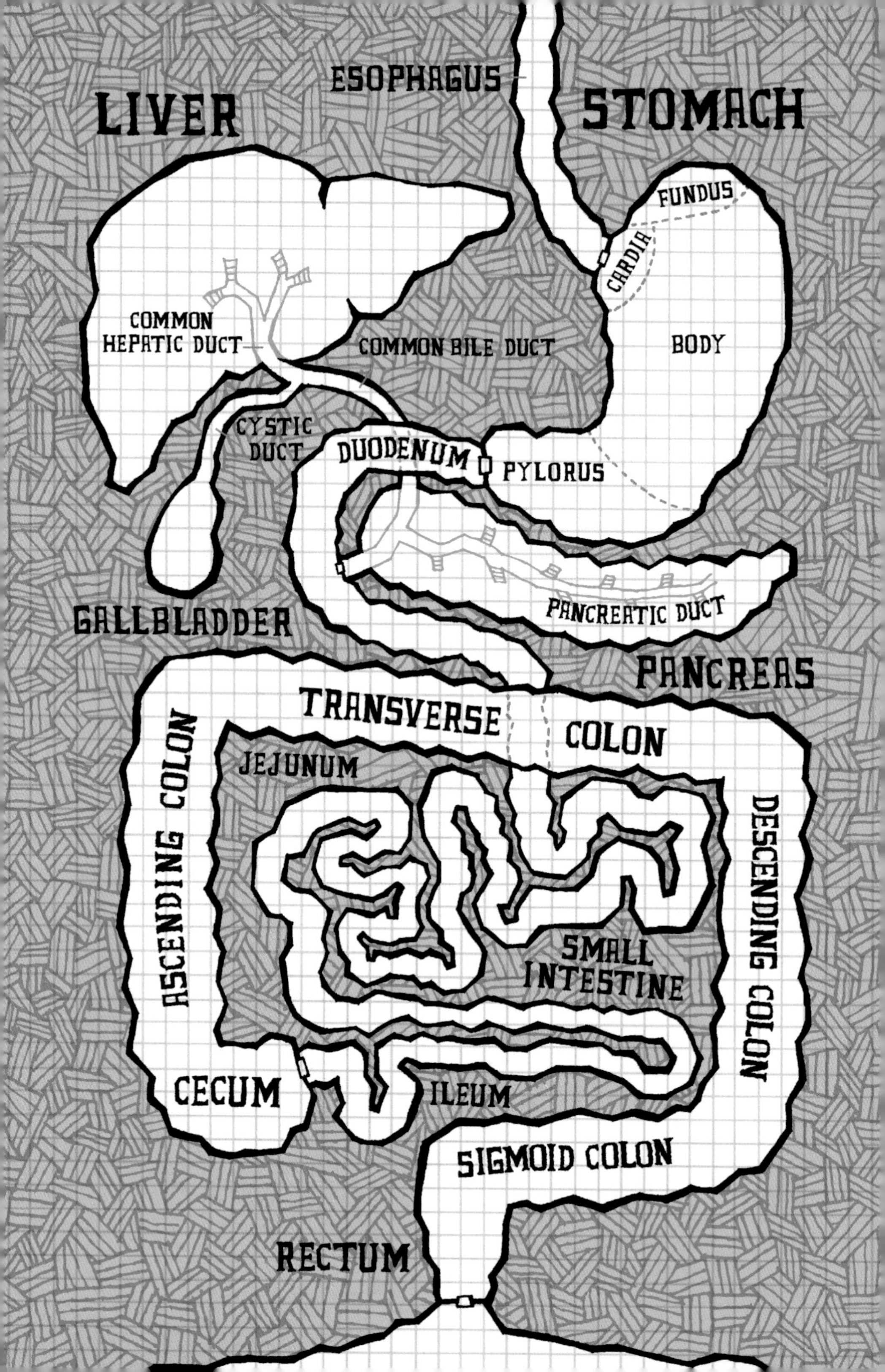
ESOPHAGUS
LIVER
STOMACH
FUNDUS
CARDIA
COMMON HEPATIC DUCT
COMMON BILE DUCT
BODY
CYSTIC DUCT
DUODENUM
PYLORUS
PANCREATIC DUCT
GALLBLADDER
PANCREAS
TRANSVERSE
COLON
ASCENDING COLON
JEJUNUM
DESCENDING COLON
SMALL INTESTINE
CECUM
ILEUM
SIGMOID COLON
RECTUM

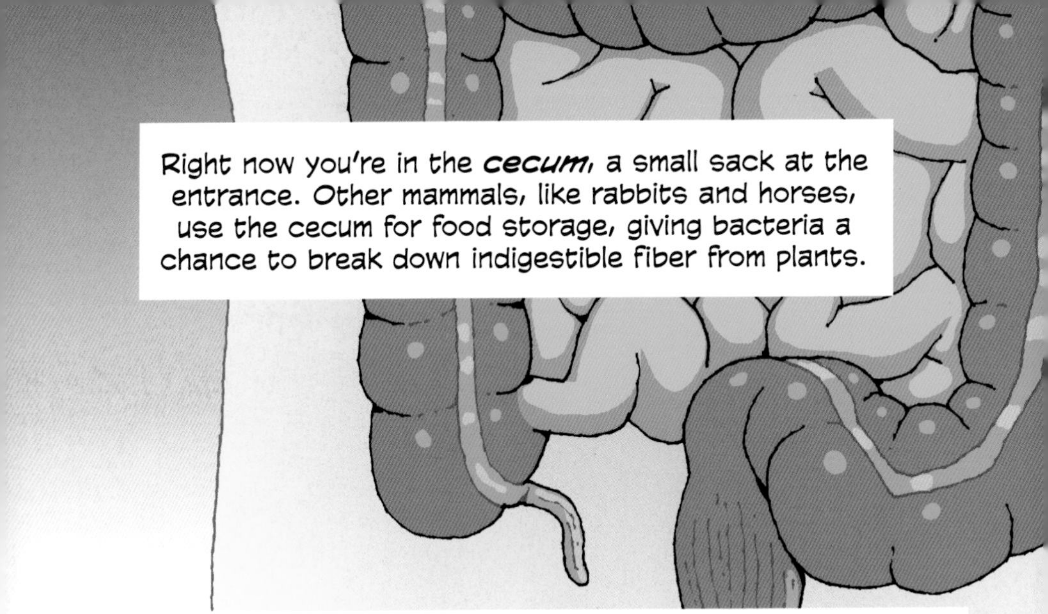
Right now you're in the ***cecum***, a small sack at the entrance. Other mammals, like rabbits and horses, use the cecum for food storage, giving bacteria a chance to break down indigestible fiber from plants.

Hanging from the cecum is a short dead-end tube called the ***appendix***. It was initially considered a ***vestigial organ*** that might have been more valuable to human ancestors.

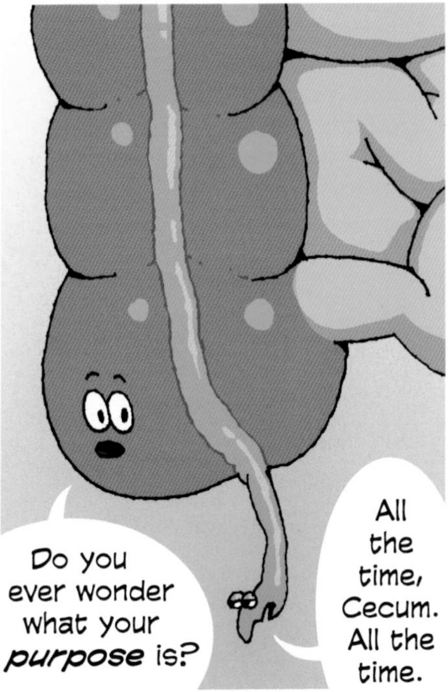
Do you ever wonder what your ***purpose*** is?
All the time, Cecum. All the time.

But it plays an important role by removing toxins and monitoring the microbes that pass above. And it may house reserves of beneficial bacteria to repopulate the intestine after a bout of diarrhea wipes them out.

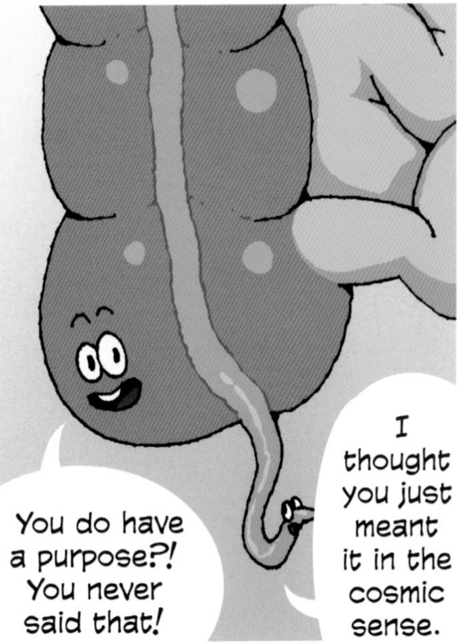
You do have a purpose?! You never said that!
I thought you just meant it in the cosmic sense.

From the cecum, chyme travels up into the colon. Most digestion and absorption is complete before food arrives, so you'll notice there are no villi on the walls. The bulges are called *haustra*.

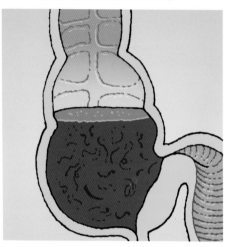

When one haustrum expands with food, bands of muscles contract, pushing the contents to the next haustrum through peristalsis.

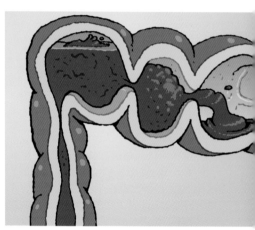

At each stop, the leftover chyme continues to be digested, allowing additional nutrients like calcium and some vitamins to be absorbed. The colon compresses the food, mixing and churning it, then absorbing the water it squeezes out.

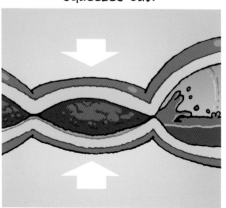

Because the food gets increasingly dried out as it passes through, there are more and more mucus-secreting *goblet cells* in the mucosa. This helps move everything along!

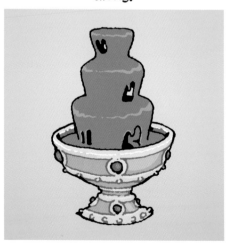

The large intestine is 1.5 meters (5 feet) long, framing the small intestine on three sides. Water absorption starts in the ***ascending colon.***

About 20 cm (8 inches) up, it takes a turn across the abdomen into a region called the ***transverse colon.*** That's where you should find Bif.

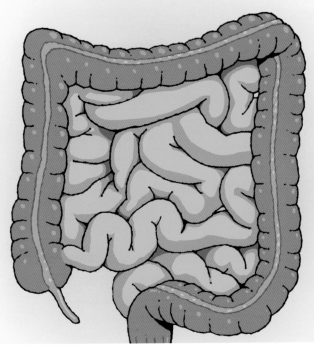

Just be patient! Contractions only occur every 25 minutes or so. And depending on the situation farther down the line, you may even get pushed backward.

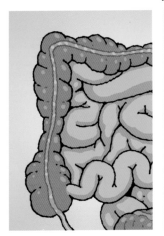

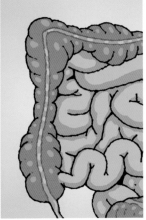

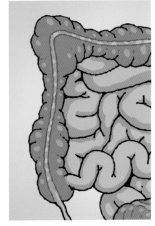

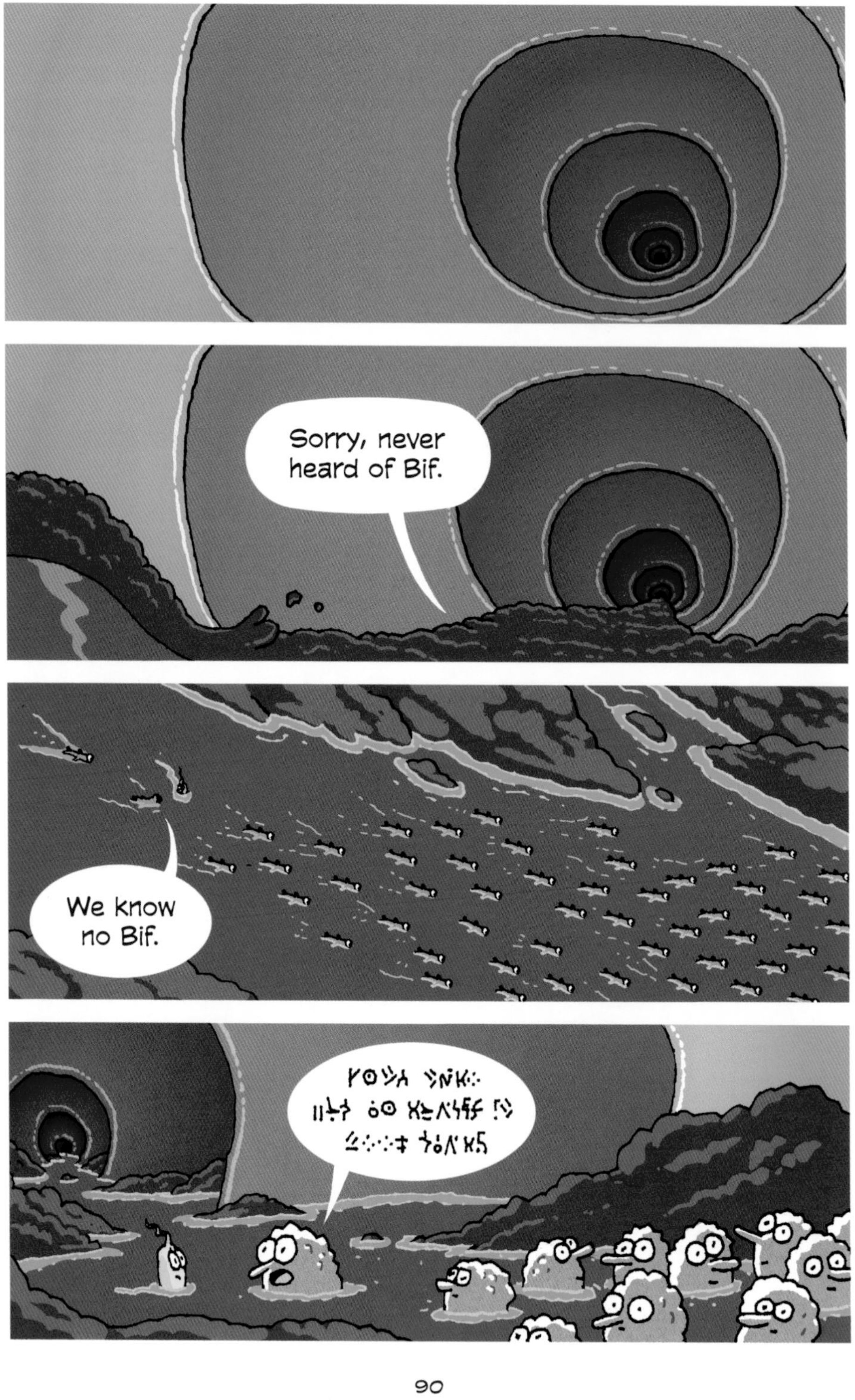
Sorry, never heard of Bif.
We know no Bif.

Wow, I've never seen so many different kinds of bacteria before. And I can't even understand some of them!

Those methanogens aren't bacteria. They're *archaea*.

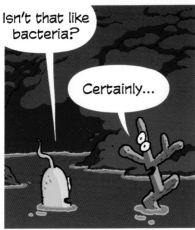

Isn't that like bacteria?

Certainly...

Both bacteria and archaea are *prokaryotes*, single-cell organisms that have no nucleus. We all come in similar shapes and sizes.

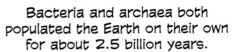

We're not so different, you and me.

We both reproduce through *binary fission*, splitting our cells into two.

Whoa!

Hey-o!

Bacteria and archaea both populated the Earth on their own for about 2.5 billion years.

Let's grow old together.

91

Then something happened. We aren't sure exactly what. But it's thought that a bacterium merged with an archaeon.

The bacterium was trapped inside, serving as an extra energy source for a new kind of cell. From this cell began the evolution of *eukaryotes,* the domain of all plants and animals.

This might get weird, but do you mind if I climb inside?

As long as no one's looking.

Never before have I felt so powerful!

Eukaryotes shall rule the world!

Calm down—

we can't do it without prokaryotes.

Eukaryotes continue to have a symbiotic relationship with prokaryotes. There are about as many microbial cells in the human body as there are human cells! Most live in the gut, where the entire population of microbes can weigh up to 2 kg (4.5 lb).

That was more information than I expected!

I always provide more than what is requested, but precisely what is required.

Bifidobacterium at your service.

Gasp!
Are you Bif?
I've searched all over for you!

Many have.

Before any bacteria arrived, the GI tract began as a layer of cells in a primitive stage of human development. The layers formed a sheet that rolled up into a tube. This tube became the GI tract, branching into all the accessory organs.

The moment our host was born, bacteria found an opening and settled down wherever they could.

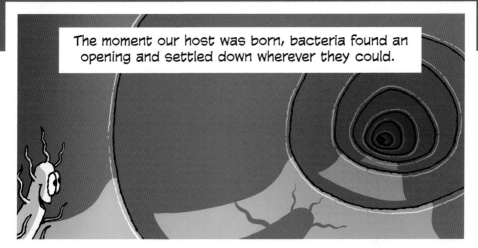

Because the immune system at that time was so primitive, it was anybody's game. Some of the bacteria living here today arrived from the mother during birth.

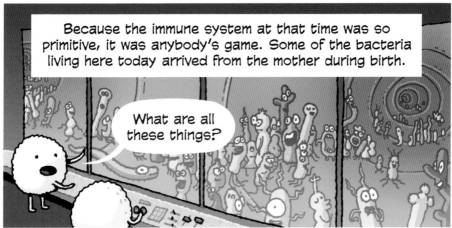

What are all these things?

Members of my clan were among those early settlers.
We migrated here from our homes outside the mother's nipple
and vaginal tract. Becoming a dominant force in the gut,
we helped digest all that fresh milk.

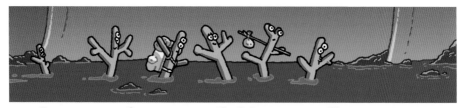

Three years of war and strife followed as different colonies
struggled for control over the same areas. Many species vanished
as soon as they arrived, not even a footnote in our history.

Those that remained learned
to live together in peace.

...and therefore confined
to the mucus membrane,
we agree to inhale and
thereby remove the
sulfur produced by you,
Prevotella, in your
pursuit of protein...

We're still around today, but our
numbers are smaller. The infant
years were our glory days.

The obsession with our
pathogenic cousins have given
bacteria a bad rap. Most of us
protect our host and keep
those same pathogens from
setting down roots.

Good bacteria break down food that the gut can't digest, producing additional nutrients. They make more sugars, fatty acids, and vitamins!

Why would anyone throw this stuff away?

They work with our immune system, training the white blood cells in the mucosa to identify outsiders and attack the threats.

A healthy diet includes not just food for the gut, but food for its bacteria as well!

When good bacteria spread out in the gut, they prevent pathogens from populating there. It's called **colonization resistance.**

Forget it. Let's just find another gut.

Dominant families have an impact on food tolerance, allergies, obesity, malnutrition, and reactions to diseases. But more than any single species, it's *diversity* that makes for the healthiest system.

Our differences make us stronger.

Most famously, our consumption of carbs causes us to release gas, producing the signature sounds associated with...

FLATULENCE

pour femme et pour homme

Farte

EXPRESS
YOURSELF

Most swallowed air gets burped out. But some of it reaches the colon and needs to go. Added to it is the gas made by bacteria that eat up all the fiber, carbohydrates, and protein they can find.

Pardon me!

If someone's small intestine can't digest sugars like lactose (from milk), or raffinose (from beans and cabbage), the bacteria in the colon have a feast.

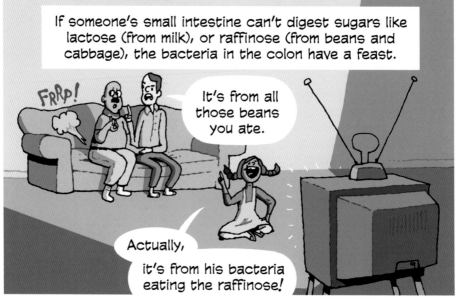

FRRp!

It's from all those beans you ate.

Actually, it's from his bacteria eating the raffinose!

Carbon dioxide is the main component of this gas—called *flatus*—but sulfur gives it the odor. The smell is usually a mixture of three main fragrance notes.

hydrogen sulfide

dimethyl sulfide

methanethiol

Hydrogen sulfide smells bad for a reason; in large concentrations, it is as lethal as cyanide.

Found this, Chief. Brussels sprouts for dinner.

Another family destroyed by farts. Such a shame.

The concentration in flatus is so low, no one actually dies from cutting the cheese. It just might *smell* like someone died.

The sulfur-producing bacteria tend to reside in the ***descending colon***, close to the end of the tube. Because it's so fresh, the most noxious flatus brings some heat with it! Anyway, that is where you are heading next.

The descending colon? But I have so much more to learn from you!

Remember colonization resistance? Look, you seem like a nice microbe. But I've been studying you this entire time and...

... I sense a pathogenic strain in you.

What? That's nonsense!

Hey, this looks like a decent place to set up shop, what do you think?

We just need to lose this *Bifidobacterium* and—

Leave now or face the full force of the bifidobacteria!

Safety in numbers, I get it!

Bif, I'm not a threat, come on! We had a thing, right? I mean...

Okay, okay...

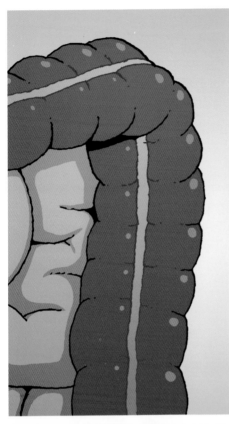

The nerve! Bif preaches about the importance of diversity, but then I get judged? Why would I damage the digestive system for my own gains? I have such respect for the institution!

Unless...he's right? No, that's impossible. I think. To clear this up, I just really need to find my...

...family.

Well, look who's finally decided to join us!

The mouth was no longer good enough for you, big shot?

It's, *uh*, great to find you all here!

I've traveled quite a distance, starting at the oral cavity, where the teeth and tongue break food into pieces while saliva kicks off the chemical digestion.

Then I got swallowed with a bolus, propelled down the esophagus by peristalsis.

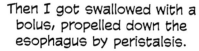

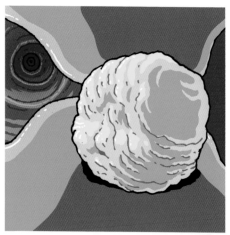

The stomach broke the bolus up physically (through grinding and churning) and chemically (by gastric acid and enzymes), turning it into chyme.

In the small intestine, villi caught the chyme and extracted the nutrients within for absorption...

...using enzymes and secretions from the liver, pancreas, and gallbladder!

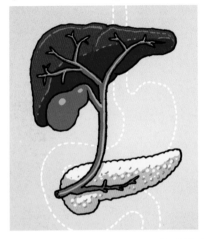

Everything slowed down in the large intestine, where the remains went through their final digestion as water, some vitamins, electrolytes, and fatty acids got absorbed.

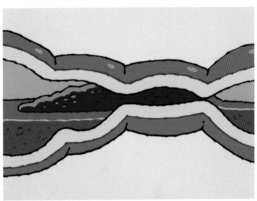

And now I'm here with you at the end! It's taken about thirty hours.

You talk like we didn't all just travel the same path to get here.

I didn't take as many notes, did *you*?

So! What is it we all do down here? Break down cellulose? Produce fatty acids? I met a bacterium who thought I was... and you're not going to believe this, *ha-ha*... it thought I was pathogenic. Which is crazy, but honestly, I don't really know my identity.

We're *Escherichia coli.*

Ah, that's a relief. *E. coli!* Most strains are harmless and many are helpful.

What do we manufacture for our host— vitamin K?

Wait, what? How is that a good thing?

We eat sugar. Sugar's a good thing.

We produce a toxin that blocks cells from making proteins and then kills them.

It's so awesome. It causes stomach cramps!

And bloody diarrhea!

I *love* sugar!

There aren't yet enough of us to have a real impact, though, so it's great you're here.

You're not going to ditch us again, right?

You're joining us?

I've only wanted to *help!*

I never thought I was a toxin-producing strain of *E. coli!*

What did you think the "E" stands for?

I didn't ditch you, and I don't know!

Education!

I want to protect my home. Not hurt it.

Look— this is not your home....

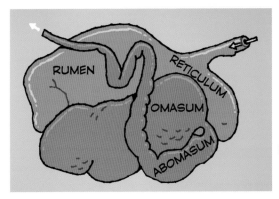

RUMEN

RETICULUM

OMASUM

ABOMASUM

Our home is in the gut of a plant-eating animal, like a cow, a sheep, or a goat. Because of their diet, these *ruminants* rely even more on bacteria than humans do. Instead of breaking down cellulose and other fiber at the end, their bacteria take it on at the start.

After passing the esophagus, food is fermented by bacteria in the first two of four chambers in the stomach, turning it into *cud*. The ruminant regurgitates the cud and chews it again.

Did you know humans can't digest grass?

Did you know birds swallow rocks to grind their food?

Fascinating.

When ruminants are slaughtered, their gut bacteria get loose. Eating undercooked meat can transfer us to a human's gut.

Their waste can also end up in fields of spinach and other produce, where even foods advertised as "prewashed" could be contaminated.

Food preparation delivers other harmful bacteria as well.
Salmonella from chicken feces get mixed around during the
slaughtering process, and then the poultry is "washed"
in tanks of warm water that spread it further.
It can also appear on the outside of their eggs.

Meat left out at room temperature can foster
the spread of *Clostridium perfringens,*
called "the cafeteria germ."

Our friends *Campylobacter* live in the intestines of
chickens and other animals. They're the most common cause
of foodborne illnesses in the industrialized world.

Don't cook your
poultry to 74° C
(165° F)!

It would mean
certain death...

...for
us!

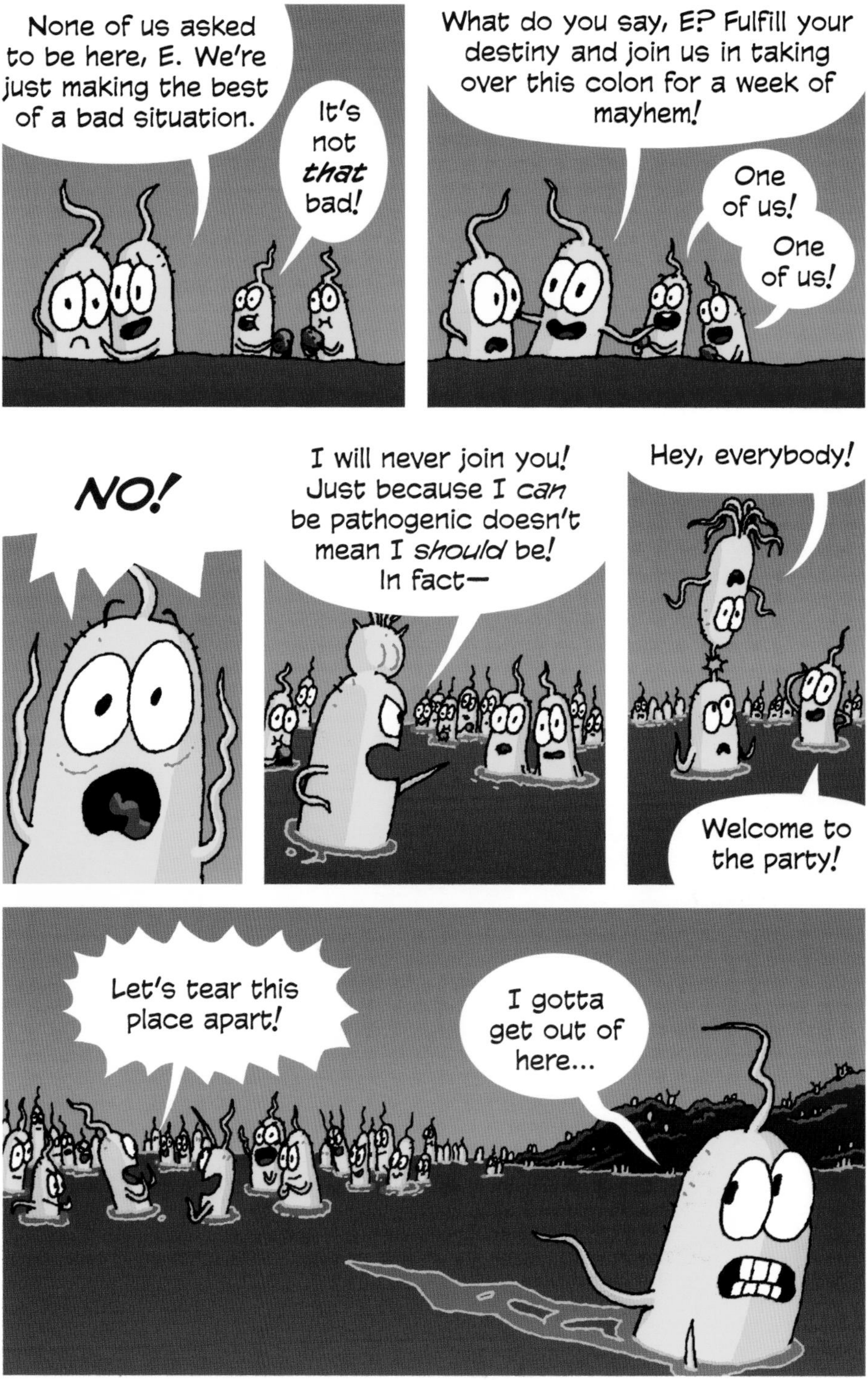

None of us asked to be here, E. We're just making the best of a bad situation.
It's not *that* bad!
What do you say, E? Fulfill your destiny and join us in taking over this colon for a week of mayhem!
One of us!
One of us!
NO!
I will never join you! Just because I *can* be pathogenic doesn't mean I *should* be! In fact—
Hey, everybody!
Welcome to the party!
Let's tear this place apart!
I gotta get out of here...

Those duplicates all came from *me!*

And I'll keep adding more if I don't do something!

PSST!

The others don't speak of this, but after all the bacteria take what they can from the chyme, some material still remains.

Two or three times a day, the large intestine gives this waste a vigorous push.

What are you saying?

There's a way out!

The waste material left after all that last-minute absorption is a semisolid substance called *feces* or stool. Feces is made mostly of dead bacteria and indigestible fiber, but also includes living bacteria, dead immune cells, medicinal remnants, food coloring, salts, secretions, and other debris!

Feces is your ticket out, and it boards in two hours.

How does it leave the body?

Through an opening called the *anus*.

At the end of the descending colon is the *sigmoid colon*. Its S-shape prevents any feces from coming out during a fart.

After you ascend the sigmoid colon...

ONE WAY

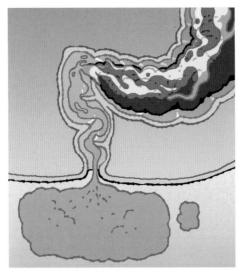

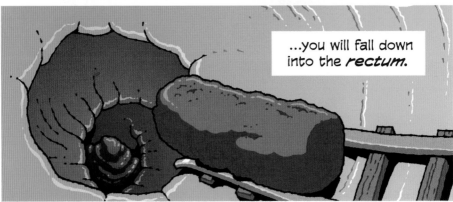

...you will fall down into the *rectum*.

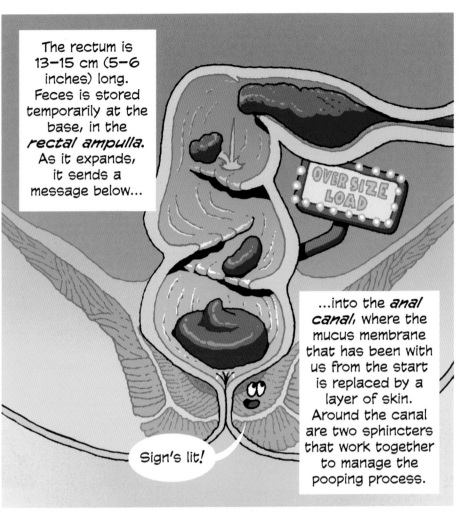

The rectum is 13–15 cm (5–6 inches) long. Feces is stored temporarily at the base, in the *rectal ampulla*. As it expands, it sends a message below...

...into the *anal canal,* where the mucus membrane that has been with us from the start is replaced by a layer of skin. Around the canal are two sphincters that work together to manage the pooping process.

Sign's lit!

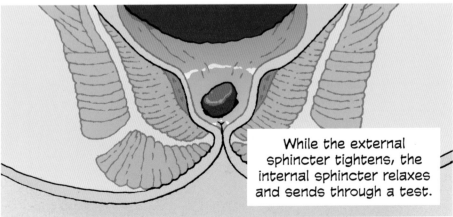

While the external sphincter tightens, the internal sphincter relaxes and sends through a test.

Sensory cells check in with the brain.

Hold it in... just a few more minutes till I get home...

Brain says no. Shut it down!

Okay, I'll send another in a little bit.

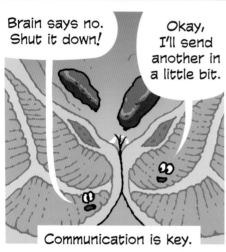

Communication is key.

But if the internal sphincter is denied too often, it may just give up.

SIGH. What's the point?

This leads to *constipation*, a condition of infrequent bowel movements (less than three times a week). Because more time in the colon means more water absorption, these poops can get hard and painful, making the problem worse.

OVERSIZE LOAD

Brain says we're on the toilet!

What's the holdup?

Maybe this time I don't feel like it.

Has the brain considered that?

I'm so glad you're helping me through all this. I don't want to stick around and cause any pain or diarrhea... whatever that is.

Vomiting is the body's way of getting unwanted stuff out of the stomach or small intestine quickly. *Diarrhea* is a way of getting it out of the large intestine. That means the body doesn't have time to absorb enough water to make a solid stool. It can be caused by bad food, infections, or stress.

You already know about vomit, right?

But this stool we're in now appears pretty healthy! Either way, this is where we depart. Thanks for letting me take you through the rectum, E.

I gotta say, poop's not such a bad place for bacteria! Friend, I think it's safe to say you've *now* learned everything you need to know about digestion!

But who knows what's on the other end of the anus!

The pleasure's mine!

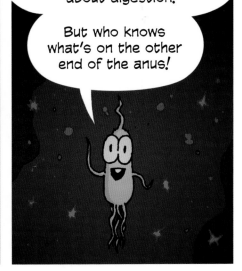

This may be goodbye, but I bet there's a lot more to learn out there!

A whole world of educational beauty!

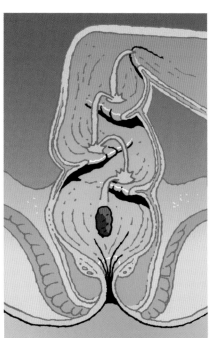

Looks good.

Let's go!

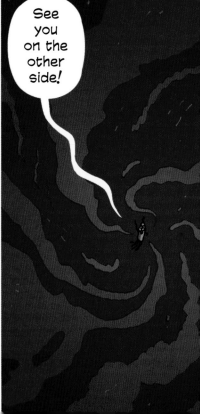

See you on the other side!

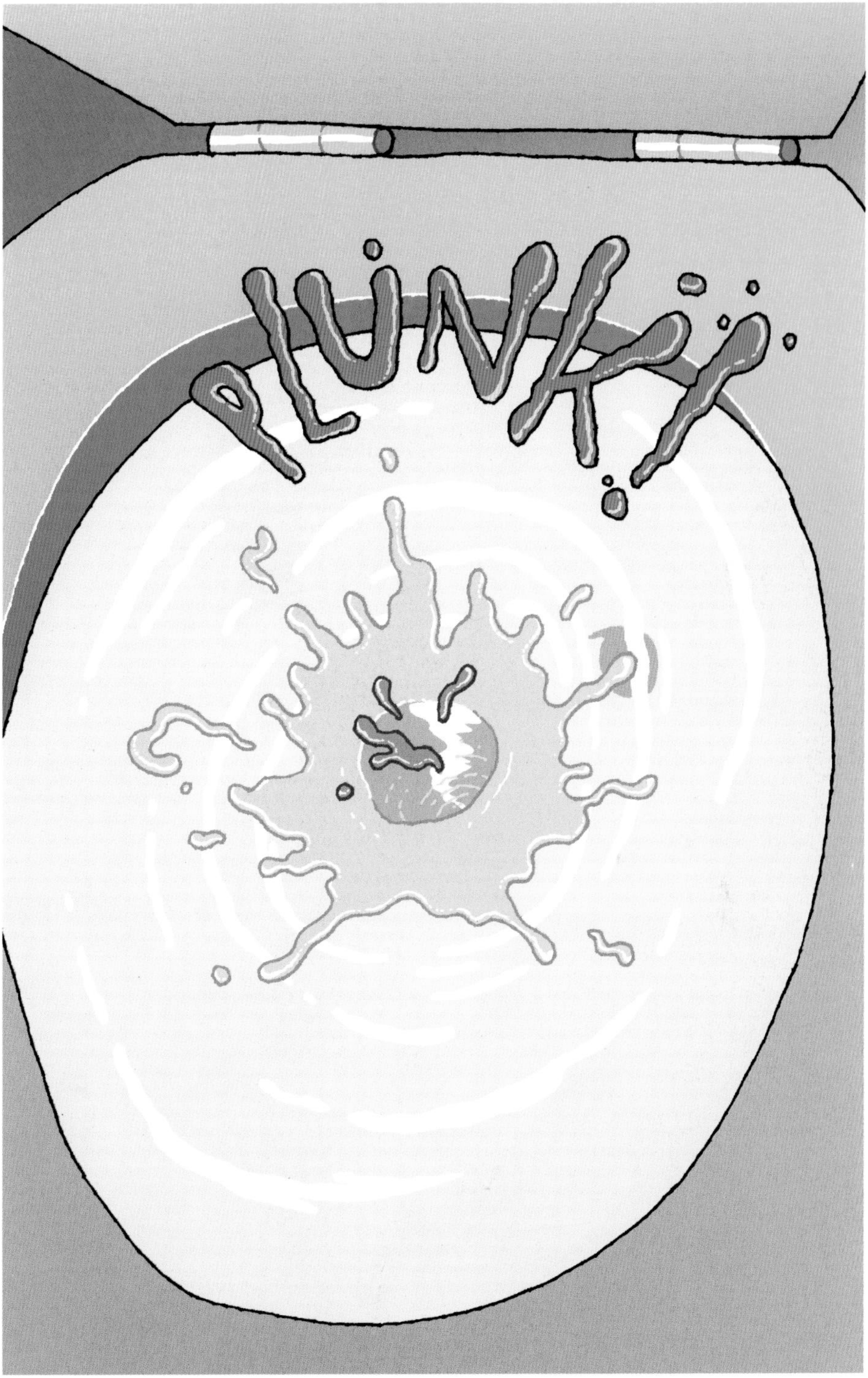
PLUNK!!

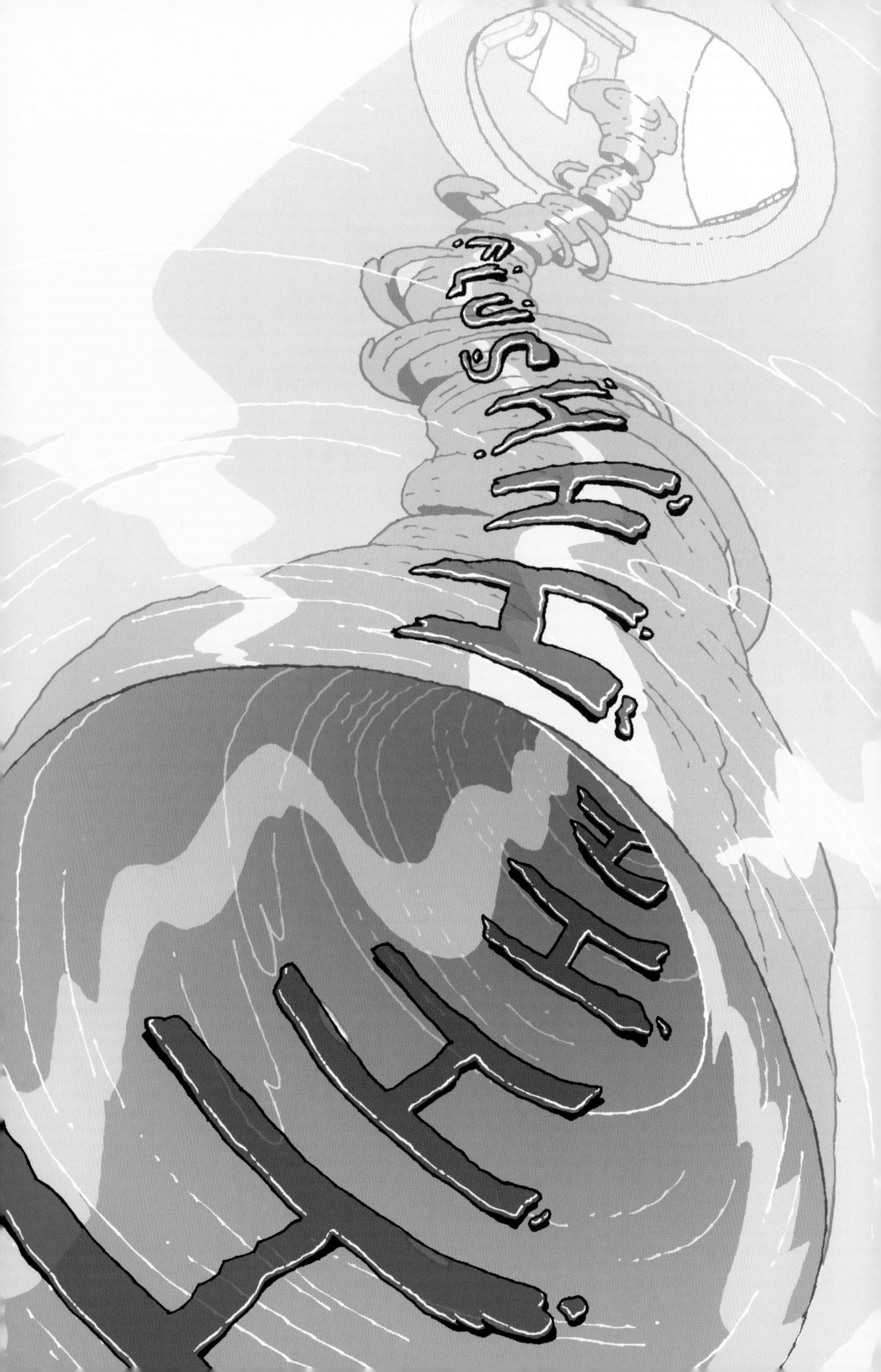
FLUSHHHHHHH

—GLOSSARY—

Amino acids
Organic compounds that combine to form proteins.

Amylase
Enzyme that breaks down starch into simple sugars.

Bacteria
Single-celled organisms that have cell walls but lack an organized nucleus.

Bile
A fluid produced by the liver and stored in the gallbladder that aids the digestion of lipids in the small intestine.

Biliary tree
A system of vessels that directs secretions from the liver, gallbladder, and pancreas through a series of ducts into the duodenum.

Binary fission
A form of asexual reproduction that splits a body into two separate parts.

Bolus
A small, rounded mass of chewed food.

Carbohydrates
Also known as sugars, nutrients that provide the body with fuel. Single or double molecules of carbohydrates are called simple sugars. More than three sugar molecules in a chain are called complex carbohydrates.

Cecum
A pouch connected to the junction of the small and large intestines.

Cholesterol
A waxy, fat-like substance that provides structure to cells.

Chyme
The thick fluid that passes from the stomach to the small intestine, consisting of gastric juices and partly digested food.

Duodenum
The first part of the small intestine just beyond the stomach, leading to the jejunum.

Endocrine
Relating to glands that secrete products directly into the bloodstream.

Enzyme
Protein that helps start or speed up a chemical reaction inside a cell.

Epiglottis
A leaf-shaped flap in the throat that keeps food from entering the windpipe and the lungs.

Eukaryote
An organism consisting of a cell or cells in which the genetic material is DNA in the form of chromosomes contained within a distinct nucleus.

Exocrine
Relating to glands that secrete products through ducts rather than directly into the blood.

Fatty acid
Source of fuel and a component of lipids.

Fiber
A complex carbohydrate that supports the leaves and stems of plants.

Flatus
Gas from the stomach or intestines, produced by swallowing air or by bacterial fermentation, expelled through the anus.

Gallbladder
Pouch that sits under the liver that stores and concentrates bile.

Gastrin
A hormone that stimulates secretion of gastric juice in response to the presence of food.

Ghrelin
Hormone secreted by gastric glands that makes you feel hungry.

Glucose
A simple sugar; the most essential source of energy in the body.

Haustra
Small pouches that give the colon its segmented appearance.

Hepatitis
An inflammation of the liver.

Ileum
The third portion of the small intestine, between the jejunum and the cecum.

Jejunum
The part of the small intestine between the duodenum and ileum.

Lactase
Enzyme that breaks down milk sugars.

Lipase
Enzyme that breaks down triglycerides into fatty acid chains.

Lipids
Also known as fats, nutrients that protect the body's organs, make hormones, and coat the outside of nerves.

Liver
Organ that detoxifies metabolites, synthesizes proteins, and produces chemicals.

Mastication
The process of chewing food.

Metabolism
The process of transforming food into energy, and using that energy to grow and sustain life.

Mucin
Protein that is a component of mucus and some saliva; helps trap bacteria.

Mucosa
Moist pink tissue that lines cavities and covers the surface of internal organs.

Pathogen
A bacterium, virus, or other microorganism that can cause disease.

Pepsin
Digestive enzyme in the stomach that breaks down proteins.

Peristalsis
The coordinated constriction and relaxation of muscles around the organs of the GI tract; creates wave-like movements that push any contents forward.

Pharynx
Passageway in the throat for food and air.

Phospholipid
Lipids that make up the structure of cell membranes.

Prokaryote
Single-celled organism without a distinct nucleus.

Protease
Enzyme that breaks down proteins into amino acids.

Proteins
Chains of amino acids that act as a building material for the body's cells.

Rectum
The final section of the large intestine, terminating at the anus.

Saliva
Watery liquid secreted into the mouth by glands, providing lubrication for chewing and swallowing, and aiding digestion.

Secretin
Hormone released into the bloodstream by the duodenum to stimulate secretion by the liver and pancreas.

Substrate
The material on which an enzyme acts.

Symbiotic
Related to close interaction between two different organisms.

Triglyceride
Lipids that provide energy to muscle cells at rest; the main component of body fat.

Villi
Finger-like projections from the surface of the small intestine.

Zymogens (proenzymes)
Inactive precursors of enzymes.

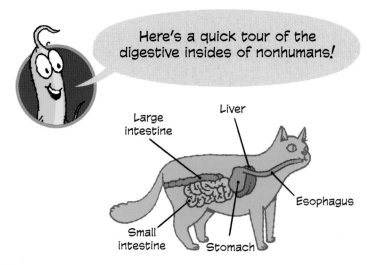

Here's a quick tour of the digestive insides of nonhumans!

CAT

Cats are carnivores! A cat's gut has a short length because it is optimized for fast digestion; a fast predator benefits from not having extra weight. Unlike dogs, cats aren't scavengers; their powerful sense of smell is particular about fresh food. Protein is what a cat's meal is all about! Their taste buds don't react to sugar, and they don't have any amylase in their saliva because they don't naturally eat a lot of carbohydrates. Keep protein and freshness in mind when considering cat food for a healthy pet!

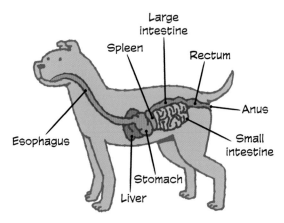

DOG

Unlike carnivores such as cats, dogs can eat a variety of foods, including meat, bones, and vegetables. They hunt but also scavenge, eating foods that would make another animal sick; their stomach acid is 100 times stronger than a human's! And a dog's mouth has 10 more teeth than a human's, featuring large canines that can tear up tough food and break down bones. A dog's digestive system has the shortest processing time of any mammal!

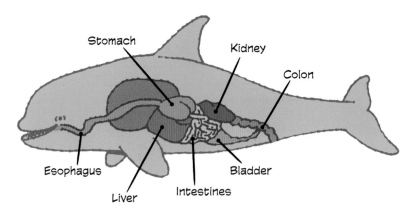

WHALE

Whales swallow their prey whole! They have a portion of the stomach similar to a rumen, which breaks down food in preparation for digestion. Some whales have sand and broken shells in there to assist. Sperm whales produce a substance in their intestines called ambergris, which helps them digest squid beaks!

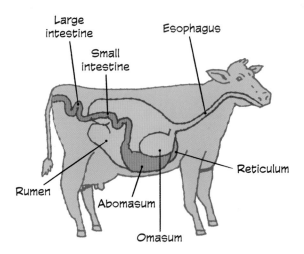

COW

Some people say a cow has four stomachs, but it actually has one stomach with four parts, all for digesting grass, hay, and other plants. Like humans, cows rely on microorganisms to help break down tough plant fibers. But because it makes up their whole diet, the microbes are front and center! The first part of the stomach is the rumen, which is full of complex microbes that get to work on the feed. If it hasn't been broken down enough, it gets regurgitated back into the mouth in a form called cud. The cow chews the cud, mixing it with saliva, softening it further before swallowing again!

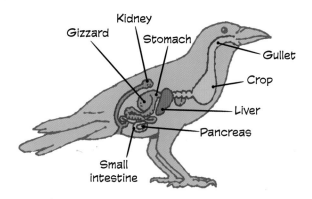

BIRD

Birds don't have teeth, so they can't chew their food! If they can't swallow their bite whole, they'll break it apart with their bill or by beating it against a branch or a rock. Most have a pouch at the end of the esophagus called a crop, which lets them store extra food they can digest and use later. A bird has a part of the stomach called a gizzard, which grinds the food down, often with the help of swallowed rocks and sand, like whales! Anything that can't safely make it through the intestines, like bones or tough husks, are often expelled from the gizzard in the form of pellets.

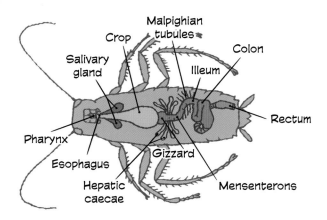

COCKROACH

Cockroaches eat all kinds of things! They have a gizzard with six sharp "teeth" that grind up the food and hairy cushions that allow tiny particles through. Insects don't have stomachs, but they do have digestive enzymes like amylase, lactase, and protease as well as microbes for digesting fiber. Thin tubes called Malpighian tubules remove waste and help regulate water. The digestive process is slow—about 33 hours from start to finish!